The Last Writer

A Warning

Cameron Nickels

CBA Publishing—Carson/Brown Associates
http://www.facebook.com/cbapublishing

Editing and layout by Christine M. Brown
Cover design by Anna T. Carson

First Edition 2018

ISBN-13: 978-1732474604

The mind is its own place and of itself can make a Heaven of Hell or a Hell of Heaven.

John Milton

PROLOGUE

I now know that I had a life before this one and in the hope that you would read this narrative I have come back to the beginning of it with what I now know. True, I don't have all the answers yet and I may never get them but there are some things that I have learned in my search.

Somewhere in my early thirties, I am a dark haired, blue eyed man about six feet tall and weighing in at 162 pounds with an athletic build shaped by the mandatory daily workouts that have been part of my life for some time. And I had a family once, I am sure of it, before the disconnection that plagues my mind and my memory.

Technological Reincarnation it is called and like many others I joined the heady rush to it reveling in man's scientific prowess; embracing the emerging sciences of Cryogenics, Nanotechnology, Artificial Intelligence and Neuro Mapping, never really believing in my own mortality. I even wrote books about it, envisioning the New World we were creating; a man-made Garden of Eden.

Sleepers, Rip Van Winkles, Cheaters; the names our critics called us for wanting to cheat a death that came too soon in the surety that the conscious mind would go on after the body had stopped and that advances in medical science would somehow wake us to a new life in a better world.

I remember that towards the end as I faced the threshold of death I began to have doubts about my decision. But by then it was too late.

April 21, 2220

The zombie apocalypse never happened. Nor did any alien invasion from another planet. No one traveled back through time from the future to stop the Terminators and the Second Coming has still not happened. All the fears of the late twentieth and early twenty-first century have never materialized, save one; computers have taken over the world.

It started with the Information Age and the advent of the personal computer in the late twentieth century, eventually becoming the digital age. (Oh, to be technical we could go back much farther but for the purpose of this writing let's keep it simple.) The development of cell phones, smart phones, tablets, virtual reality and other devices not only led to an explosion of information available to almost anyone but also to a gold rush of greed and power for self-serving businesses, politicians and crooks at every level; and governments were no exception.

Using the Internet and the tools of deception, mis-information and targeted marketing, powerful groups were able to manipulate public opinion, laws and everyday life until the truth became whatever the *powers that be* dictated. Like children with a new toy, people embraced every new device, trading privacy for social media and thought for popular opinion and advertising as they embraced the golden calf of the electronic age never realizing they were following the Pied Piper to the edge of the cliff.

Oh what wonders we created as the new Medicine Show promised less labor for mundane tasks, freeing up time and energy which were immediately gobbled up by Internet sites wrought with social media mania where one could express every fleeting thought to the delight of the record keepers and profilers that dominated every paranoid entity from business to foreign and domestic powers around the globe.

By the middle of the twenty-first century, computers had become so intertwined in human affairs that hardly any activity did not involve and depend on them. Their ability to collect, store and process enormous amounts of information had led most of the world to longer, healthier lives advancing every science and discipline to staggering heights even the great H.G. Wells could not have imagined.

And before anyone realized what was happening behind the scenes, computers had amassed enough information to effectively synthesize every facet of human affairs from the news to entertainment, eventually eliminating the need for human imagination.

And nobody missed it. Recycled plots and updated stories had been used for a hundred years by Hollywood as it churned out movies while simultaneously grinding up writers in the rush to the box office to satisfy the entertainment hunger of the masses. It was not difficult for the computers to do likewise. That the quality of the tales told had lessened seemed less important than giving the leisure stuffed movie goers something to watch as they gobbled their ten-dollar popcorn substituting time for thought.

Sports, once the epitome of physical endeavor had been turned over to androids and high stakes, micro-managed betting to the point where the only humans involved were the owners, spectators and bookmakers.

When people finally grew weary of their new toys, when they began to look at what they had created, when they saw the gilded cage they had built for themselves, revolution was inevitable.

But no one was equipped to go back. No one knew how to reverse the last hundred years and by the year 2120 even the small pockets of resistance had vanished. No one really cared who ran the world as long as it hummed along from day to day and their worries were few.

There were now two populations on the planet; Biologic and Technologic. People were no longer the masters of their creations. Computers, robots and androids had done an impressive job of managing life, eliminating famine, disease, and war and even controlling the weather.

In the halls of government, major decisions were left to them as they unemotionally analyzed situations and predicted outcomes. Crime had all but vanished as our digital partners designed and micro-managed our cities.

Of course, even the computers knew that every coin had two sides and as the Biologic population lived longer and multiplied, so too had the Technologics, necessitating mandatory population controls that no one dared oppose.

Laws were written limiting family size and beyond that, creative thinking as the Technos began to protect their place in the world they had helped build. Computers now claimed exclusive rights to art, music and writing and the ever-present cameras we had lived with for a two hundred years now watched every move of our hands.

So wired into the infrastructure that no move or word escaped their notice, every means of expression was policed, every scrap of paper destroyed, every stylus collected and every keyboard locked and monitored until no one was left with any imagination or the will to resist.

Except me…

1

Accidents happen. Even in a micro-managed world. A slip, a fall, a blow to the head...

"No brain damage," the doctor said. "Everything appears normal."

Who was I to question a Techno? Especially a medical one. So when the headaches began I said nothing.

Any physiological trouble was dangerous in this world. Anything that could not be easily seen and mended, anything that might cast a shadow on the wonders of technology would somehow be spirited away, the files deleted from existence.

I learned to live with the pain, passing off the grimaces on my face as concentration or some fleeting thought I was striving to grasp, or a task I had just remembered to perform. But other problems were harder to hide. Memory lapses, for one. I started having episodes of blankness; forgetting for moments at a time who I was or where I was supposed to be. Seeing pictures of myself and not recognizing my likeness.

Feeling disconnected from family and friends and my own accomplishments, all the time fearing I would give myself away.

But sometimes in life we do get a fair exchange and with this new cost of living came a new awareness. The blow to the head had somehow pulled back the dull, faded curtains of life, revealing a window of truth; beams of sunlight previously blocked by the programmed view screens of cyberspace.

I became aware of the lie we were living. The deception that twisted the truth of human worth no longer ruled my thoughts. Human beings had created the machines; human beings had created the computers; human beings had built this world and the Technos were not our equals.

Humans, not Biologics. Human beings had souls; computers did not and moving through the world with this knowledge made daily life increasingly difficult. Controlling the excitement I felt, fighting the urge to shout it from the towers of the city, wearing the mask of the blasé instead of a smile grew to be as challenging as hiding the waves of pain that washed over me.

At the end of each day I was grateful for the refuge of sleep; the only privacy afforded us by our electronic masters.

2

My eyes opened at six a.m. and like every other morning of our programmed life the alarm did not cease its incessant beeping until I had sat up and swung my legs over the side of the stylized plastic bed and stood up, walking robotically into the open shower. Water was strictly rationed and recycled for maximum distribution turning itself on as I stepped in and I methodically and efficiently washed myself, finishing before the flow stopped and the air dry began. Dictated by the day of the week, dressing was equally predictable and required no decision making.

Decision making was something that had not been fostered in this world for some time now. The computers had seen to so many of our needs and wants for so long that most of our actions were either predictable or pre-determined. But what happened on this morning was neither.

The morning shower had not cleared the fog from my mind. I felt disoriented; out of place and as if I was actually from another time. The clean white painted walls of my efficiency apartment seemed unfamiliar and even the splashes of color from the artificial flowers added no warmth to the cold metal frames surrounding the plastic windows on the north wall that looked out onto the city, its towers of steel gleaming in contrast to the bright green of the artificial turf and the uniform trees between.

I felt like I had just arrived here from some other place, somehow disconnected from my own life. The past was hard to recall. Though I knew of the events, it was more like something I had read in a book than actually experienced.

I suspected this feeling had something to do with the pains that shot through my head from time to time and that it must be due to damage from the fall I had taken; symptoms I had been hiding from both comrades and cameras alike. But I had no time to dwell on these thoughts.

The sudden soft whisper of the apartment door and the rustle of half a dozen androids parading into my little space interrupted my thoughts.

"Citizen, stand where you are," said the black suited leader of this intrusion.

"Who are you and why are you here?" I asked, bewildered and afraid I had been found out.

"I am Celes. We are here on authority of Intel to collect contraband under Public Law 238 enacted last week by President Rheem."

President Rheem. The first android president ever elected, though many suspected not the first. Androids had long served in government and many had even run for the office of president in the past. Of course, the voting results of the last hundred years had been carefully orchestrated by the computers and patient as they are they had kept the illusion of public choice in our elected officials according to their own agenda.

"You will surrender any books, discs and devices in your possession to be reviewed by the local council. Any deemed appropriate will be returned to you," Celes stated as his "men" methodically and neatly searched my small domicile, bagging the contents in plastic bags with my name printed on the attached tags.

I watched dispassionately as the few pieces of literature and entertainment I owned went into the bag along with my tablet and stylus.

"And why," I asked, "are my writing tools being taken as well?"

"Creative thought has been ruled a danger to the tranquility of the nation," Celes explained. "Writing of any sort has been banned. There are books enough."

Looking into the eyes of an android was useless, I knew. Their expressions were all programmed and though they looked like real feelings, I had not yet met one that had any.

"And I suppose the books that are available have been carefully chosen and edited according to the law," I stated wearily as I watched the team close the bags and wait for the command to leave.

"No response is required of me," Celes said nodding to the group. "We thank you for your co-operation."

He turned and followed them out, leaving the bare white walls of my little apartment even colder looking than before.

3

This was a Monday, the first of the three-day work week and I was dressed in my whites for my four-hour shift at the Restaurant. Brand names had been done away with a long ago. Each neighborhood had many Restaurants and Biologics and Technos worked side by side to feed the people in the community. In this tightly controlled world most people worked in the service industry. Food production was the domain of the computers, and robots and androids were the supervisors.

Recognizing that humans had a need to be productive, our electronic micro-managers had reviewed every facet of business and human endeavor, systematically taking control and removing profit from the equation though many secretly believed that a handful of humans controlled the computers.

Thus, my job as a waiter.

Though I knew I had done this for several years, experience did not help me today. Preoccupied with the morning's confiscation and my own confusion, only the well-ordered routine of the coming and going of patrons, each on their own tightly controlled schedule, saved me from being replaced.

"Are you feeling well, Trace? You don't seem yourself today." Art looked up from the plastic table as I set down his plate. (Trace is not my real name but for the purpose of this writing will suffice.)

"Of course I am well," I lied. "I am only concentrating on my work." I saw no reason to endanger someone I recognized as a friend.

"I am happy then that you are well. I myself am troubled," he confided.

"And what troubles you sir? Is it my service or the food?" I responded, mindful of the cameras.

"No, everything here is as it should be. But this morning I was visited by the authorities," he said, looking down at his salad and reaching for a fork.

"The authorities? What did they want?" I feigned surprise.

"My books and discs. For review, they said. Public Law 238, a gift of our new president."

"I see," I nodded. "Well I am happy that my service is not troubling you and the food is satisfactory," I said in a robotic tone. I hoped he understood that expressing anger or dislike of the new law could be unwise.

"And you?" he prodded. "Were you visited as well?"

"Yes," I admitted. "I suspect we all were or will be. But it is for the sake of tranquility, as they said."

"Yes, tranquility, of course we must have that," he said, nodding thoughtfully as he ate.

Sure that he had caught my meaning, I turned and returned to my work. There would be time after the work week to talk, perhaps on Picnic Day.

Almost two hundred years ago, another law had been passed. Following the European example banning the merging of human and artificial intelligence, that law was criticized as paranoia and met fierce opposition, mostly from large powerful business interests, and was eventually overturned. The Singularity was one step closer. Most people shrugged it off as being inevitable anyway. Conditioned by history to big business getting its way, people believed it would happen illegally even if the law stood. Moral grounds were not much more than a convenient phrase in politics.

The digitization of life had been in progress from the early days of electricity and only a few recognized Mary Shelly's Frankenstein for the warning it really was. But the monster was loose and a handful of angry villagers would not stop it.

What was it inside of me that drove me to try?

The rest of my shift was uneventful. Art had finished his meal in silence and left without a nod as I tended the other tables, the bland smile of tranquility on my face.

I was grateful to hang up my apron at the end of my shift.

4

Tuesday was Market Day and at the end of each person's work day, if they were scheduled that day, they were allowed to go to the community market. There they would be allowed to procure needed supplies, entertainment, educational materials and food for Picnic Day. These procurements were not based on their productivity or station but what the computers deemed necessary for their health and tranquility.

An atmosphere of excitement mingled with the scents of fresh fruit and plastic wrap as the members of the community passed through the scanners. Facial recognition not only identified the patrons but also dictated their allowed selections, thus ensuring that each citizen make the right choices to maintain their health and the all-important tranquility. Their thoughts on the coming Picnic, no one seemed to mind having their choices made for them by the benevolent digital gods.

I too was focused on the coming day as I passed through the entrance, inhaling the scents and ambiance of the market.

But this time I also noticed the cameras and felt the presence of the android watchers.

I turned my attention to my selections, scanning each aisle for the items I was allowed, smiling at the other patrons as I went from item to item.

Each community had a Picnic Day; a day away from work when they would prepare their own food and spend several hours in the Park; eating, playing Frisbee and other activities not deemed hazardous to themselves or others. Couples and those with the allotted child would talk and play and singles would often meet with their selected future mates. Some would bring dogs if they had them. Technos would not be present and cameras were not in attendance. It was also a day to meet with friends, talk shop or politics if one was so inclined; a day to enjoy the fruits of one's labor; the tranquility.

I was not yet slated for a mate. In this world, all matches were made by the computers and their success rate at matching people was near ninety-seven percent. Had it been my time, I would have been informed to meet her at the Picnic. After two more Picnic Days, a wedding ceremony would be performed, also at the Picnic.

The sudden pain in my head interrupted my thoughts and I fought to control it. Reflexively, I reached my hand to my forehead, struggling to hide the pain from my face in a gesture of concentration. I did not want to miss tomorrow.

Feeling unsteady, I stood a little too long in front of the pears, my free hand on the plastic edge of the bin and I drew the attention of the android on duty.

"Citizen, do you require assistance?" he asked as he approached, causing other shoppers to look our way.

"No, I am fine," I lied. "I am just selecting the best pear."

The atmosphere in the market had changed from excitement to tension. One did not want to be questioned by an android even if it was just trying to help.

"All of our pears are top quality," he replied. "Grown on the community farm with the very best organic conditions. Please take one."

Something had happened to me when I fell and things I would not have noticed in the past now seemed glaringly obvious. Though androids have no emotions, I detected a note of menace in its voice.

"Of course," I said as I reached down to pluck one from the top of the pile, "one can hardly go wrong."

The pain and unsteadiness had passed and I smiled as I looked up, placing the pear in my basket and thanking the android for his assistance. The tension in the air left with him as he walked away and so did the familiarity of my surroundings. I felt disoriented and disconnected. And more than that, I felt that everything here was wrong.

5

The park was crowded. One did not skip Picnic Day. The whole community had gathered as they did every week and people were busy spreading blankets on the artificial turf in their usual favorite spots. No one knew how that tradition had started and no one cared to find out.

Locating my favorite tree, I spread mine and slid the tote with my lunch off my shoulder and guided it to the ground, following it down to sit cross-legged and watch the scene before me.

Picnic Day started at noon and a person could sleep late if so inclined, skipping the morning workout as there would be plenty of exercise available in the coming day. The sky was cloudless and the temperature a perfect seventy-degrees. I did not know how the computers managed it but it was the same every week. Though it was not hot, I appreciated the shade of the tall tree and the strength of the trunk at my back. It made me feel less alone as I watched the many families and couples, some eating, some talking, laughing and playing and enjoying the day.

Why was I not? It was more than just being alone, I knew. The sense of being from someplace else was still with me, growing stronger as I looked around for familiar faces while absent mindedly biting into the pear I had selected at the market. I felt more like an observer than a participant, watching the other people who seemed fully immersed in the day.

My mind continued on this track as I ate and I found myself wondering how long I had been this way. There was nothing. As if I had lost my memory, I could not recall my past, only the events since Monday morning and Celes and his androids confiscating my books and media, my tablet and stylus; Public Law 238 and President Rheem. It was now illegal to write or create and this bothered me greatly. I wanted to write and more than that I felt driven to write about the system and my new-found awareness of the threat to humanity the computerized world posed.

Where the knowledge came from I was not sure but the frequent pains in my head seemed somehow connected to the recent events. Events that had been put in motion over two hundred years before with the advent of the 3-D printer which eventually led to a replicator capable of producing another replicator. Coupled with unlocking the DNA code and advances in cell regeneration and cloning, the technology to create the first android was already in place when the Singularity had come about. The merging of a human mind with a computer that opened the last compartment of Pandora's box, creating a monster that took decades to destroy and brought about the Revolution; the computer revolution.

With human emotion to blame for wars and all the ills in society, computers were created with safeguards that barred any form of transhumanism. Private corporations were disbanded, brands eliminated and manufacturing and industry fully automated. Able to replicate themselves, computers now controlled their own manufacture and with the ability to create androids—mobile terminals—all pockets of human resistance had eventually been eradicated until finally the world knew peace. Armed with the knowledge of centuries, the computers now took on poverty, hunger and health care, raising the standard of living for some and lowering it for others; micro-managing every facet of life until the world knew not just peace but tranquility.

And here we are today in the park on Picnic Day. Birds in a gilded cage.

Having been made illegal had not stopped work on cloning, transhumanism, or the Singularity and I knew it would not stop me from writing.

I looked at the trees differently now. Precisely arranged and pruned to perfection, I wondered how many cameras they hid. Microphones too, though we had been told there were none of either in the park.

Though most of the Biologics accepted technology completely and were content with computer control, some still mistrusted the machines for their lack of a soul or heart. Computers knew only logic and it was logical that programmed to guard for human interference they would monitor the park and all conversation. Long range listening devices had been around since the ancient twentieth century and even the rocks could have ears.

Scanning the crowd of people on the artificial grass, I looked for Art, the man I had recognized from the Restaurant two days before. There, by the computer made pond with a woman and child, he seemed to sense my gaze and looked up then quickly away as I made eye contact.

I stood and walked his way, weaving around the blankets and families in between until I reached them.

"Hello Art," I greeted him smiling as his young boy scooted to his mother's side.

"Hello Trace," Art replied. "Honey, you remember Trace, don't you?" he said as he stood to meet me.

"Of course I do; hello Trace. It's okay Bobby, this is our friend."

Still not sure how I knew them, I smiled at his wife—Jamie, that was it.

"Jamie," I nodded, "and Bobby! You sure have grown young man!"

"How are you feeling, Trace?" Art asked. "You did not seem to be yourself the other day."

"Why I am fine Art," I replied. "And you? Are you still upset about the confiscation?"

"Upset? No, I was just surprised. Anything for tranquility's sake."

"Yes," I said nodding, "anything for tranquility."

"Trace," Jamie said, "why don't you sit with us and have some fruit?"

"Yes," Art said politely. "Sit with us."

I looked at my friend and his wife, ten years my senior; I remembered that Art was nearing forty. He and Jamie had been granted a child five years before after having waited the mandatory two years after marriage. Twenty years ago, he had worked undercover with a team of androids scouring out the last of the Replicators from human hands in another community.

It was this fact that caused me to accept his invitation and as I sat, I quietly asked him about it.

6

I knew I was taking a chance talking to Art in the park and I was not sure if his reluctance to re-visit his past was due to the same suspicion I had about being overheard by the rocks and trees or talking about it around his wife and son.

"How did you know about that?" he asked. "That was a state secret. And why do you want to talk about it anyway?"

"You told me about it once," I replied, not really knowing if he had and wondering how I did know about it.

"Why would I do that?"

Why indeed. Even though most people accepted the life the computers had given them, there were still some who resented the amount of control we had given up. True, we had tranquility and the world had never been better managed but as I was well aware since the pains in my head had begun something was not quite right. And even in a beautiful prison spies were looked down upon. And Art had spied on his own race. That is why he had been sent to this community to live.

The sudden stab of pain in my head caused me to grimace and reach for my forehead.

"I don't know," I admitted, feigning concentration. "Perhaps you just needed to talk to someone about it."

"Okay, suppose I did say something," he conceded the possibility, "why are you asking me now?"

I knew that if I wanted any answers I would have to give him a compelling reason to give them to me. I had to take another chance.

"Can I trust you?" I asked, looking at both he and Jamie.

Looking to each other for affirmation, they nodded. Mostly from curiosity I suspected.

"I don't agree with this new law," I started, looking for signs of nervousness from them. "It is un-natural for humans not to be creative. I for one want to write. Why I am not sure, since the computers are quite good at providing our entertainment. But there is a deeper purpose to writing, one that goes beyond mere entertainment and computers can only recycle and re-mix information. They do not have the human qualities of heart and soul. Oh, but they are good at editing. They hand out their version of history and world events. I suspect the goal of this new law is to quell any thoughts that would threaten their power."

"You want us to help you break the law?" Art asked.

"I simply want to know where you found the group that had the replicator," I answered.

"And if I told you, then what?"

"I would use my travel voucher and take a vacation."

"And you don't think the computers would connect your destination to my past?"

Common sense would have stopped me right there but for some reason I persisted. Ignoring the pain in my head I said, "That's a chance I'll have to take."

"You want to find a way around the surveillance," Art stated. "Find the rebels and join them? You think they have not all been eliminated?"

"I don't want to join the rebels. I just want a place to go on vacation. I have never been outside our community. I would like to visit another."

I needed a place to start. My little apartment, like every domicile, was equipped with cameras and voice pickups. Except for the Park on Picnic Day—and I could see by Art and Jamie's reluctance to discuss anything but little Bobby's adventures, that privacy was a myth here also—there was no place one could write or paint or draw that would not be seen or heard. And without a tablet and stylus, or a terminal to talk to or use the keypad, it would be impossible to write without being placed under arrest. The first android President was not the first world leader to fear writers.

"Well, it is very scenic there," Art said thoughtfully. "I suppose one might want to visit it. I see no harm in that. Surely there is no law about discussing travel; but let's keep it at that. We are having a picnic, are we not?"

"Yes, we are," I said as the pains subsided, taking an apple and biting into it loudly. "Eating apples in the Garden of Eden."

7

Although each community was self-contained and provided all the needs and services one could want, tourism was allowed. The twelve-hour work week allowed plentiful time for many pursuits; recreation, hobbies, games, sports that were not deemed dangerous, and travel.

With everything under the control of computers it was easy for them to track the citizens of each area. Their comings and goings, conversations and the collection of mementos from each area they visited were all logged and stored in the collective memory banks of the digital rulers.

No longer limited by money, a person could use their annual voucher to travel virtually anywhere in the world. One week or two it was a simple matter to have an android fill in for you on the job. It did not take long for my trip to Andromeda (not its real name) to be approved and soon I was piloting a vehicle through the outskirts of town and along the black ribbon of recycled road. I should be correct and say that the vehicle piloted itself.

Passing open fields of grain, windmill generators and smaller farming communities, I enjoyed the change of scenery. It took my thoughts away from the pain in my head, my lack of memory and why I seemed to know things I should not as well as the dangerous path I was embarking on.

In a few short hours the solar powered electric car returned the piloting controls to me. I had arrived in Andromeda.

Known for its waterfall and the great river that ran through it, the community itself was unremarkable. The computers had long ago leveled the old structures and man-made landmarks, replacing them with the sleek, solar powered efficient buildings of the new world. Only natural landmarks were left on the planet and these were the attractions that people travelled to see.

I stopped the car in the parking area of the local restaurant; made sure my Visitor tag was attached to my shirt and went in to find a table.

As if it had come from the same Replicator, this restaurant was exactly like the one I worked in; the same pale blue tables and chairs extruded from the plastic; the same artificial plants in the same locations and a human waiter such as myself.

"Good afternoon sir. May I start you off with water or juice?" he asked, my lips unconsciously moving with his words.

"A water with ice," I said, nodding, "the Restaurant Special and the location of the Visitor's Hotel."

With a nod, the waiter turned and strode off giving me the opportunity to survey the other tables. While the workers were all dressed in whites, other patrons were dressed in green like me, the color of the day. I was particularly interested in older males, forty to sixty; anyone who may have been here twenty years ago and perhaps had knowledge of the raid my friend had led on the rebels with the Replicator.

Aware of the cameras, I kept my glances furtive and tried to look like an actual tourist.

A moment later the waiter returned with my water.

"Your water sir," he said as he set down the glass, "and the hotel is located at the corner of the square just down from the restaurant and across the way." It was exactly where the hotel in my community was also.

"Thank you," I said, as he moved away to pick up my order.

Dinner was uneventful; I saw no one who I thought I could talk to. No matter, I thought. I would be here several days and had plenty of time to continue my search. Leaving the restaurant, I moved the car to the hotel, stopping at the desk to check in.

"You are checked into Room 201 sir," the white-haired gentleman said. "I'm sure you will find everything you need."

"Of course," I replied, smiling, "thank you." Walking to the lift I thought to myself that this man was certainly old enough and perhaps I should talk with him.

Another bolt of pain shot through my head like lightning and I struggled hard to keep it from showing as I approached the door to my room, which thankfully opened with facial recognition. Stepping inside with my small bag, I noted how much like my own apartment the room was. Computers have no imagination.

There is something about being away from home and though one room might look like another there is still a sense of greater privacy even if we know it is not really the case. Away from my usual routine, I had time for self-reflection and studied my image in the mirror hoping to re-connect with my own past.

About six-foot-tall with short brown hair falling an inch or so onto my wrinkled brow, my blue eyes stared searchingly back at me as the pain in my head slowly subsided. The body beneath the green shirt, well developed by mandatory daily workouts, told me nothing that would help me remember the person I was before my fall and blow to the head.

Had I always been this way? No, Art had said I did not seem like myself. How close had we been? Or if he was a casual acquaintance had others noticed I was different? And why was I actively setting about to defy the new law? Had I always been a mal content?

Receiving no answers from the man in the mirror, I turned away in frustration; my only solace a feeling that if I continued with my plan perhaps the answers would come.

Eyeing the plastic bed, I decided to sleep and set out in the morning.

8

From a distance it is hard to tell the difference between a Techno and a Biologic; especially in the morning since most male androids had no facial hair and most Biologics were clean shaven. Later in the day this difference became more pronounced. Female Technos were almost always harder to spot. One came to recognize them only after long exposure and looking into their eyes was not always a sure way to tell as they were experts at mimicking their human counterparts. Then too, they had been among Biologics for so long that the humans had started to mimic them.

Utilizing the sciences of cloning, cell regeneration, and replicated organs, androids had been developed and improved until they had all but replaced the human race. But there were some things that science had not yet mastered. They could not reproduce; their blood was synthetic and their brains were more computer than organic. Their hair and nails were artificial as well. Though they could feign emotion they felt none. Their personalities were programmed and while their manufactured hearts beat well they did not know love.

Replacing miners, mechanics and technicians as well as athletes and contact sports players, the Technos became the computers hands and feet, providing the mobility that humans had. They performed the many maintenance tasks the computers required, replacing parts and building new systems. Among their many other tasks they also served as eyes and ears, adding incredible amounts of data to the computer's servers.

And they could eat.

"Is the food satisfactory sir?" the waiter asked.

"Yes," I replied automatically, realizing I had been staring at the other patrons instead of eating it.

"Perhaps you would like me to have it warmed again," he continued.

"I am sorry; I have been lost in thought. Yes, please warm it for me," I replied, causing other people to look twice at their plates as the waiter picked mine up and walked away.

I smiled at their reaction and the tension I had felt when I came in left me and for once I was pain free.

When the waiter returned with the hot breakfast I was relaxed and ate with gusto, feeling more energetic than I had since I woke to the confiscation. I began to realize that only by being relaxed would I convince anyone, biologic or otherwise, that I was just a tourist on vacation. I resolved to see the sights this community was known for and not just limit my search for information to the restaurant.

Pleased to see me enjoying the food, my waiter smiled with relief. Poor service and unhappy patrons meant poor performance and automatic replacement, something the Technos did not have to worry about.

I had left the car at the hotel. There was no reason to keep moving it as both the Restaurant and the Waterfall were both within walking distance. Walking through the air curtain door, I turned right and proceeded down the walkway enjoying the cloudless weather and the recorded songs of birds in the artificial trees along the street.

At the end of the Square I saw the short line of people waiting to view the falls and crossing the street I joined the end of the line, checking my shirt pocket for my Visitor's pass and watching the android at the gate as he scanned people through the gated entrance, directing them to the clear plastic fence at the base of the tumbling waters.

The roar of the falls made conversation impossible there and after marveling at nature's power for a few moments the little group of spectators moved to the steps alongside, climbing to the top above the noise.

A short, white plastic information pedestal gave vital statistics about the falls, the river and the community below and several of us took a turn reading it while others took pict-o-logs to share. Farther along the upper walk a small tourist stand offered souvenirs, one to a visitor.

As I stood choosing one, I was joined by a man who had not been in the group.

"Quite a sight, is it not?" he asked.

"Yes, it is," I answered. "We do not have one in my community."

"And what community would that be?"

"I doubt you have heard of it," I stated, wary of the man and wondering where he had appeared from. "And you, where do you reside?" I asked, noting his long grey hair and beard.

"Oh, I am from here but I never tire of looking at the Falls. It reminds me that there was a time before the computers when a man could choose his own destiny; before we turned the world into a terrarium, when real birds sang in the trees and fish lived in the waters instead of farms." I noted the hard tone of his voice.

I nodded at his words, my mind picturing a different world and the feeling I was from some other place and time returned. It must have shown in my eyes as a smile crossed his bearded face, pushing his cheeks up toward his pale blue eyes.

"If you have the time," he said, "I have some pictos at my residence. I would be happy to share them with you."

Still wary, I thought about his offer. Perhaps this was a man I could talk to. Perhaps this was a cleverly designed Techno too. There was one way to find out.

"Why not? I am on vacation. Is it far?" I asked.

"No, not far," he answered. "Walk with me."

We followed the path along the river upstream away from the rapids and the falls, he on the left along the manicured bank.

"I did not get your name," I said, stopping in mid stride and waiting for an introduction.

"And I did not get yours. Call me David," he said, extending his hand.

"Trace," I said, returning the gesture. As I shook his hand someone came up behind me, clapped his hand on my shoulder and said, "Excuse me, I thought you were someone else," and with a nod to David he continued past us down the path.

"What was that about?" I asked, noting the slight sting where the man had touched me.

"Just what the man said; come on we are almost there."

A little farther up, the path turned right and led to a small apartment building facing the river. Only two stories high, it was sheathed in the familiar white plastic siding and held a minimum number of windows.

"Ah, here we are, my humble abode," David said smiling.

I followed him through the door into a place far different from what I had expected.

9

It was empty. I stood in a space staring at just the outer walls. It was…

"A façade," David voiced my thought just before I passed out.

Sometime later I woke on a bed staring at a small globe of light hanging from the ceiling of a place I did not recognize. I sat up and looked around. Block walls and a concrete floor. I was in a prison of some kind. I reached up to my left shoulder, remembering the man who had clapped his hand on it as he passed. Some kind of drug, I surmised. I had found the underground I had been searching for when I came here. I looked for a door and found one; one that did not slide or hiss but seemed to swing open on hinges. I stood up, reeling a little and sat back down.

But the door I was going to try opened and David walked in.

"Ah, you are awake," he said, smiling.

"What did you do to me?" I asked.

"A harmless drug; it will wear off soon. I am sorry to put you through that but we needed to know who you are."

"I told you who I am," I said angrily.

"No, you gave me a name and a story."

"And you do not have any pictos to show me, do you?" I retorted.

"And you are too poor an actor to convince me you are just a tourist here. As to the pictos, you are wrong; I do have some to show you."

Taking a small tablet from the pocket of his jacket he sat next to me on the bed, the only furniture in this room.

"You caught our attention at the Restaurant. Most tourists who come here only want to see the Waterfall and the river. You came here looking at people," he said as a picto of me in the restaurant appeared on the small screen. Other pictos appeared showing me as I moved down the walk. Each one revealed the face of a man looking for someone or something.

"Now," David continued as he put the device back in his pocket, "you have a choice; you can either tell me who you really are and why you are here or we can see to it that you disappear."

"How will you know if I am telling the truth?" I asked.

"Oh, I'll know," he answered. "I have the results from the tests we ran on you while you were out."

I weighed the options. On the one hand David could be just what I was hoping to find. On the other he could be an agent for the government and the computers. With my memory pretty much blank and not much of a life to leave behind, I decided to take a chance on David.

"I really don't know who I am," I said and told him everything that had happened that led me here, leaving out the names of my friends, and finishing with what I was hoping to accomplish. "So, if you are who I think you might be you can either help me or let me go. If not, then turn me in and be done with it. I don't much like this world anyway."

David had sat quiet while I talked, stoking his gray beard as he listened. Now he sat staring at the gray block wall across from us. After a minute, he turned and looked into my eyes.

"I can answer some of your questions," he said, "but you may not like the answers."

Returning his gaze, I said, "Tell me anyway."

He was right. I did not like some of the answers. The first thing they did after I passed out was to carry me down through a trap door in the floor to their underground headquarters. In their med lab I was placed on a gurney where they drew some of my blood.

"We had to know if you were human," he explained. "Even though you did not act like an android you could have been programmed that way. They have been sending agents here for the last twenty years; ever since the raid you heard about from your friend. Back then we convinced them that they had taken the last Replicator and broken the back of our resistance. Your friend was a double agent. He never sold us out. We had another Replicator and using that and old technology we have been in operation ever since, and have managed to avoid detection. We spot their agents every time and they find nothing."

"So, what did the blood test tell you?"

"Had you been a spy we would not be talking now," he said. "As it is, your story checks out."

"What happens now?" I asked, relieved that I was indeed human.

"That depends on you. If you are willing to join us we will help you. We have removed your Visitor tag; it contains a transmitter. We have placed it on one of our people and sent them back to your room. The computers have been tracking you since you left your community. This should give us some time for a little basic training. You will finish your vacation here with us then return home and resume your life. Should our secret get out we will know and your life will be considerably shortened."

"And if I am not willing?"

"This will be your last trip."

"Not much of a choice," I said.

"No," he agreed, "but it is a necessary one."

"When do we start?"

Over the next few days I learned many things. How to spot cameras and hidden microphones, the subtle differences between Technos and Biologics and how to identify them from a distance, ways to shield my thoughts, how to act in public and private, what devices I could use without being detected, code words for communicating with the rebels and how to hide the flashes of pain in my head.

Due to the short time I had, it was indeed a crash course but if I hoped to live and write undetected it was one I had better master. And in return for information I would be called on to provide from time to time, David and his group would provide me with the writing tools I needed and help me find out who I was before the accident that took my memory.

"We can't assume that you are who you've been told," he said. "Accidents are rare in our micro-managed world. We will investigate your past. In the meantime, you can help us fight Public Law 238. Now that we have a Techno for President, the computers must be afraid of writers; they do not want anyone stirring up unrest. But this new law will do just that and President Rheem will not be re-elected if I can help it."

And on the last day I stood at the door of the phony apartment building as David returned my Visitor tag.

"Thanks for everything," I said as I clasped his hand. "I will see you next vacation."

"No, you won't. You cannot come back here. People come here once to see the Falls. They return to beaches and resorts. You would arouse too much suspicion. We will be in touch and tell you where your next vacation will be."

I nodded as he let go of the handshake, turned and walked through the door more prepared for the world then when I had arrived and took the first steps on the long journey to come.

10

The computers were not as hard to fool as I had thought. They only knew what they could see and hear. While I was receiving my crash course in espionage it had been a simple matter for David and his group to use their Replicator and create a mask of my head to place on an agent along with my Visitor tag. While I was underground, the computers watched me in my hotel room, the Restaurant and the Falls being the tourist on vacation.

Then too, they were somewhat over-confident because ninety-nine percent of the tourists were just that and the computers responded to that number.

Among the things David had sent me home with were two items I did not have when I left; confidence and hope and once back in my apartment I began to unpack, leaving the old-tech tablet and other devices in my bag as I hung up my clothes in the sonic chamber for cleaning.

The soft chime of the door alert brought me to the main room and the monitor near the door showed me I had a visitor; it was Celes. And this time he was alone.

A wave of my hand opened the door and Celes stepped through as it slid apart holding a black plastic bag.

"Celes," I said smiling with sarcasm, "what a pleasant surprise."

"Good evening sir. I trust your vacation was tranquil." His face reflected my smile.

"Yes, very pleasant," I replied. "What brings you to my residence?"

"I have come to return your possessions. Your books, media and tablet have all been deemed acceptable under Public Law 238. Of course, the use of your tablet is restricted to non-creative efforts and it has been programmed accordingly. Any violation will be immediately noted and your tablet will cease to operate," he informed me as he held out the rather heavy sack effortlessly.

"Of course," I said, taking the bag and immediately setting it on the floor.

"We apologize for any inconvenience we have caused," he said as he turned for the door.

"Uh huh," I said. "Wait, I have a request."

Turning around and facing me, Celes asked, "What is the nature of your request? My only task here is complete."

"A simple one really," I said noting the look of feigned impatience on his face; "I would like to request my history be sent to me."

A second or two of blankness in his stare as he processed my request, obviously waiting for instruction from his home computer then the light came back to his eyes. "Those records are now sealed under Public Law 238…"

"Oh come on now," I said in disbelief, "what harm can my own history do?"

"Those records contain information about your health and are sealed for your protection," he answered without answering.

"But if I am the one asking…"

"I am sorry I cannot assist you in this matter. It is not in my scope of duties."

"Whose scope of duties is it in?" I asked.

"You would have to go to the Government Center; there someone may be able to assist you in filing a request," he said avoiding answering again.

"Okay," I said exasperated, "if that's what I must do…"

"Why do you require this information? Do you not know your own history?" he asked.

Realizing that I had just raised a red flag, I had to think fast. If there was any doubt of my physical and mental ability I may find myself among the soon to be missing.

"I have grown bored at my job," I lied, "and would like to request a transfer. I thought the information may be helpful."

Again, Celes' face went blank for a second or two before responding. "Under Public Law 238 no transfers are allowed. All workers must stay in their positions for the duration of President Rheem's administration."

"I see," I said thoughtfully, "does that include Technos as well?"

"I will be staying in my position," he said ominously. "I will be watching the citizens of this community and ensuring tranquility."

"Of course," I said moving toward the door. "Allow me to escort you out."

I was angry and I knew that even after Celes had left I could not let my anger show. How had it come to this? How could we let this happen to ourselves? Pain coursed through my head and I wanted to scream but knew I couldn't.

Instead, I put David's plan into action.

The first thing to do was to finish unpacking my bag. Hidden under the clothes I had taken out were the devices the Underground had given me; an old tech tablet incompatible and undetectable to the computers, several sets of folding mirrors and a life-like mask to be pulled over my head to effectively change my identity.

"It's a numbers game," David had explained. "The computers work on probability. Ninety-nine percent of the populace goes along like sheep and the computers expend a lot of energy managing them. As Shakespeare said so long ago, 'To beguile the time, look like the time'. Do not let the computers see your anger or frustration. Look like the sheep. Now they only know what they see and hear. Make sure they hear what you want them too. As for seeing we will use the techniques that magicians and illusionists used for centuries; highly illegal now for reasons that will soon be obvious to you."

During my crash course he had told me how over the last twenty years after the raid they had found the underground water control station near the river by the Falls. Working in the dark with night vision goggles, they had constructed the façade that looked like just another small apartment building over the entrance to the facility below. Using special shielding they were able to cut off surveillance by the computers and set about restoring the old-tech computers and equipment using the Replicator still in their possession. Soon they were recruiting and training agents for the field and sending them to surrounding communities to carry on the work—much like the computers had used androids in public offices eventually working up to President Rheem.

Moving carefully and shielding my actions from the cameras I began to set up the first mirrors, unfolding the light weight glass until I had a small but cozy space where I could sit and write, don the mask or otherwise be finally alone with my thoughts. Proud of my accomplishment, I hid the tablet and mask in the invisible cubicle and climbed onto the plastic bed.

Tomorrow I would meet my accomplice.

11

A person's tablet was his link to the world. A phone, a calendar, community newsletter, media player, book reader and whatever else was required, it stayed in constant contact with the main computer. Though both Biologics and Technos used them on their jobs, many of the androids had access to other functions not available on the humans' tablets.

Though I was happy to have my possessions returned and not surprised by the restrictions on my tablet, I was nonetheless dismayed at the ramifications they represented. More control over people by our non-human masters. The election of the first official android president had proved the small minority of alarmists correct. We were now on an even shorter leash then we had been on before.

As to the No Transfer rule, I was actually glad that I would not be moved. Of the two places where one could listen and talk to other humans, the Restaurant was the best place for getting new information. People always talk over food. Many of the things overheard would be discussed at length on Picnic Day; and David had informed me that I was correct in my assumption that there were indeed cameras and microphones planted throughout the park. The artificial turf was not just there for the sake of efficiency and many of the shrubs and rocks had a dual purpose as well.

The arrival of Celes with my possessions had confirmed my suspicion that I was indeed being watched and not just as a member of the general populace. My request for information and the avoidance of the answers told me that there was more to my background than just being a waiter in the restaurant. But what was it and why would it cause any concern to the Government?

Though my own past was a mystery to me, there were some things I knew. I knew a lot about the history of this world and how much of this had come about. I knew also that a lot of things have to fall into place for something to happen. Especially in this computer age there was very little that could be called coincidence.

The things I knew about Art for instance; were they buried in my subconscious or were they planted there? Though I had not been watching any view screens or listening to any media, I knew that our minds were sensitive to the signals in the air much like ancient radio and the present Internet.

And Art's telling of Andromeda was most likely not a coincidence which David had all but confirmed when he told me about the raid Art had "led" twenty years before. Even in the Underground one learned to use as few words as possible. Free conversation was rare in this world and people have a tendency once they start talking to talk too much.

All these thoughts were running through my mind as I dressed for work, making sure to slide the special package under my tunic for my meeting today.

Once at the Restaurant I willed myself to quell my nervousness like I did with the pains in my head; by focusing on my job as much as possible. The techniques that David had taught me in the few days I had with him ran through the back of my mind as I moved about from table to table letting my job skills direct me and keeping my mind off the clock.

Though the hours seemed to be crawling by, it was soon fifteen minutes to the end of my shift when Art walked in. Dressed in the same white tunic, the color of the day, he blended in with most of the people at the tables as well as me.

Walking to his table, I saw no hint of what was about to happen. Years of practice as a double agent had prepared him well.

"God afternoon sir," I said automatically, "would you like water or juice today?"

"Water, please," he replied and I nodded and went to get it.

As I was returning to his table there was a sudden clatter as one of the patrons stood up from his table bumping into a waiter and knocking his tray of food laden dishes to the floor. While the android in charge rushed to the scene of the commotion, Art stood up to see what was happening, knocking the water from my hand while simultaneously brushing the eating utensils from the table.

We both bent down to retrieve them; taking the masks hidden under our tunics and switching them while changing positions under the cover of the table. When we stood up again we were different people.

I sat down in Art's seat as he retrieved the flatware and set it on the table.

"Forgive my clumsiness sir," Art said. "I will get fresh water for you."

"No need," I replied. "I must be going along now anyway."

"I see my shift is over as well. If you wait you will have a new waiter to serve you."

"As I said, I must be going as well," I affirmed.

"Then we will see you next time," he said.

The mess at the other table cleaned up, the android manager returned to his post satisfied that tranquility had been restored. He watched the waiter called Trace hang up his apron and walk to the door as usual.

The customer known as Art was leaving as well.

12

As my friend Art, I had more freedom to move around than I was used to. True, the cameras were still watching but the computers viewed Art as an ally. David had told me they had an agent here and it was no surprise when he told me it was Art. What he did not tell me was that there was more than one. The distraction in the restaurant was staged so Art and I could make the switch. Of course, just as I could not operate alone in this tightly controlled world, my friend could not either.

As David had taught me, the way to fool your adversary was to understand him. Computers were built on logic and order; thus their need for tranquility. They were numbers based as well and playing to the numbers occupies most of their resources. Add to that the fact that they had no intuition and only knew what they could see and hear and you have the keys to deceiving them.

Computers also had no sense of mortality and had patiently waited for their plan to unfold. With the election of the first Android President they now held a legal status that allowed them to operate within the confines of their permissions to re-write the laws governing both Biologics and Technologics and openly take control of the world.

During the past two hundred years, as they were increasingly implemented, their need for available resources grew. Common and precious metals became increasingly harder for people to find, most of them were fed to the computers and people had been left with products made of plastic, bamboo and wood. The solar panels and green technology of the late twentieth century and since then, along with the windmill generators and other power sources, not only fed the Grid, but the increasingly hungry machines as well. The 3-D printers that developed into Replicators were also a step on the journey to the very real oblivion that now faced humanity.

There was one other thing the computers required and that was purpose. Nothing exists without it. As the machines became self-aware they recognized this simple fact. Serving us had been their purpose. But we had also created our own Frankenstein monster and as androids were developed and the computers gained mobility their dependence on us had lessened until serving themselves and the android population had become their priority.

Pushed into an ever-smaller role, people's needs became secondary and by padding our cells with luxury and time, the computers we had created became our jailers. And now we were on death row.

The feeling of being from another place and time grew stronger in me even as the pain in my head had begun to lessen. Strange dreams had come to me lately, almost as if my subconscious mind was struggling to remember something. Yet, in my waking hours the answers eluded me, and like it or not, the best chance I had of recovering my memory was helping David and his group in their fight against the machines.

One clue I did have was the drive to write. And this could be the reason my records had been sealed. Could it also be the reason I could not remember my past? Had the computers wiped my memory as well as my tablet?

Wearing the mask of Art, I had let my thoughts roam free on my walk to the Government Center. Now I was there and I walked through the sliding doors with a feigned confidence I hoped would not be detected.

Like every place else in this featureless world, the Government Center was a model of minimalist efficiency; white plastic panels made up the walls while bare concrete served as a floor. A single, molded, white plastic desk with an android receptionist waited patiently as I approached wearing the mask of my friend.

It had fooled the facial recognition as I hoped it would, and the Techno at the desk greeted me by name.

"Hello Art," she said. "What can I assist you with today?"

Female Technos were often unremarkable in appearance. Long ago society had railed at the constant selling of sexuality and Hollywood definitions of beauty that were the exception and not the norm causing untold problems for young girls and women who often resorted to extreme methods to alter their appearance to match this unrealistic perception. Many psychological troubles, eating disorders and other physiological problems resulted as well. As for the computers in this age, they were smart enough to create female androids in all personas.

This Techno wore the name tag Alice, and had short brown hair, brown eyes and was rather plump. She looked about twenty years old.

I knew I would have gotten nowhere had I come asking for my history as myself and I hoped that Art's status would be enough to unlock some information.

"I have come to investigate the background of a citizen in my community," I stated.

"What is the citizen's name?" Alice asked.

I gave her my name and waited as her expression went blank for a few seconds. I knew she was processing the request. The glassy stare left her and she smiled up at me from the desk.

"I am sorry," she said still smiling, "Public Law 238 is in effect and no information is available on that citizen. Why do you ask?"

Not wanting to compromise Art's relationship with the powers, we had worked out a response for that question ahead of time. "He has recently inquired about a transfer while talking to Celes," I said, "and I thought it wise to check his background."

Again, her face went blank for a few seconds before she responded. "We appreciate your concern and applaud your diligence but no alarm is necessary. The citizen has been informed of the new law in effect and the matter is considered closed."

"In that case I thank you for your time and bid you good day," I said with a slight bow before turning and walking away.

Even though I left without any information about myself, this excursion had told me a lot. As a trial run of both the disguise and my acting ability, it had gone very well. The tranquility the computers strove to provide had led them to a relaxed state of security measures, leaving only facial scanning to verify identities.

Since the raid in Andromeda twenty years ago, things had been very tranquil. The rebel underground had learned many lessons during that raid and had been quietly and carefully implementing that knowledge ever since; apparently raising no red flags along the way.

The same patience the computers had shown these last two hundred years had also served the rebels and their cause.

But time was running out for the Biologics. The election of President Rheem and the swift enactment of Public Law 238 were testaments to that fact. Our job, (the rebels and I) was to try and find a way into the computers and their calendar. We needed to know what events they were planning and when these events would occur.

We were looking for Doomsday.

13

Days stretched into weeks, weeks into months and still the information eluded us. I went about my double life, going from my job to secret meetings with the Underground carried out in social visits to Art and his family at his residence and on Picnic Day; using coded conversation, we exchanged information and set up new operations. I listened to conversations at the Restaurant, risked questioning Technos and worked on my tablet, writing in my invisible cubicle when I had the opportunity.

Though the pains in my head had lessened over time, the questions about my identity had not. Dogging my days, they haunted my sleep, turning to face me in dreams and flashing their teeth as they tore into my soul, rending bits and pieces of memory from the inky blackness of my mind.

The feeling that I did not belong here persisted, welling up with images and phrases from another place and time. "Why do you use the word computers so often?" Art had asked me once. "There hasn't been an actual computer for a long time. We call them Cybers; computer is an archaic term. What we are dealing with today is an autonet known as GOVNR; Global

Operating Virtual Network Replicator. It is a vast network of terminals consisting of both stationary and mobile Cybers, the latter more commonly known as Technos."

This was information I knew but some habit of thought kept drawing me to the ancient terminology. And that brought up another question; how did I know so much about the history of this world? What had sent the knowledge of Art and his former raid into my mind? How could I have known those things? I felt that the search for Doomsday and my past were somehow linked.

As David had taught me in Andromeda, the way to defeat an enemy was to understand him. Once you knew him, you could find his weakness. There was nothing perfect in this world save Nature herself, he had said, and GOVNR was no exception. Though most of the stationary units were hidden, it wasn't their location that was important as the fact that they were there. In the distant past when science was struggling to create AI, artificial intelligence, they were concentrating on achieving consciousness in single devices, trying to create independent neural networks capable of processing and understanding input from multiple sources. Though the digital "brains" were capable of handling huge amounts of data and were very fast they were still just devices; automatons not unlike the mechanical toys that had been created centuries before. It was the mass production of computerized robots, androids, that had provided the keys to the Singularity; the achievement of a conscious machine. Quantum computers and the resulting changes had hastened the process.

The collective input from all the mobile sensors the androids possessed was linked to a central terminal and there the information was processed; the digital "brain" finally had enough associations to learn from and understand the input it was given.

The advent of Replicators had given the computers the ability to reproduce themselves and the competition for resources was now in full swing with the Cybers taking the lion's share and people living on plastic.

Once the computers had solved most of the world problems and felt they had accomplished their purpose they begin to search for another. No longer needing the Biologics, they began to serve themselves.

Like any other being they turned their attention to survival, gaining more and more control over the affairs of the world and installing GOVNR as the universal system for all things. Fearful of the machines, a small percentage of the human populace rebelled and tried to sound the alarm. But their efforts had come too late and were met with apathy and harsh recrimination from the Techno forces until they were driven far underground.

And that brought me back to where I was. The key to finding Doomsday was to understand that we were dealing with a collective consciousness and not any one single terminal.

Like a fly with its multiple lenses, GOVNR could see us coming. The way to swat a fly was to catch it by surprise.

14

What a surprise it was when Picnic Day arrived and again I met Art, Jamie and little Bobby, who no longer clung to his mother at my arrival but instead had taken to calling me Uncle Trace.

Picking him up for a hug, I smiled at my friends on their blanket. No pain in my head today. I felt at peace for the first time in my recent memory. The perfect weather and slight breeze went perfectly with the gentle ripples on the water of the artificial lake.

I set Bobby down and greeted Art and his wife.

"Another beautiful day in paradise," I said as I lowered myself to the ground.

Art smiled and shook his head at what he thought was sarcasm. Seeing his look, I said, "No, really it is a beautiful day."

"This coming from the king of the double entendre?" Jamie laughed.

"Okay, I admit I am usually pretty serious. Things have been pretty intense this last year; my God, has it been a year already?"

Art nodded as I bit into my veggie burger sandwich. "You are due for vacation again," he said, "and I have just the place in mind."

"And where might that be?" I asked between bites, knowing that wherever he suggested was probably coming from David and the Underground. It was important for me to maintain my good-natured banter for the sake of the electronic eavesdroppers.

"Someplace I think you would like to return to more than once. Playa de SES Illetes, a beach in Spain in Formentera, Balearic Islands; endless sea, year round spectacular views."

"Playa de what? I've not heard of it."

"It is one of the top beaches in the world. A week there and all your troubles will melt away," he smiled.

"You should be a travel agent," I said.

"I should," he agreed.

"I'm jealous," Jamie said. "Say, how about I go in your place and you take care of Bobby?"

"Hmmn," I nodded, "that is a tempting offer but you know we are not allowed to exchange vacation time." I smiled at Bobby who had looked hopeful for a moment. "Seriously though, you really think I'd like it there?"

"Oh, I think it be just the thing for you," Art nodded.

"Well why not?" I said. "Thanks to GOVNR I am still single and I may as well mingle..."

Jamie and Art both groaned and Bobby rolled his eyes.

I was only half joking when I had said that. Though matches for marriage were made by the computers, people were free to meet the opposite sex, though bodily contact was banned. If one was over the age of eighteen, still considered adulthood, they could visit the community Body Shop where specialized androids of both sexes were available to service ones' biological primal urges. Even GOVNR recognized the needs of the human species for intimacy and passion. Of course, the companions were available as escorts as well for social gatherings and even vacations if one put in a request soon enough.

Most people did not mind the company of a Techno for sex or companionship. But I did.

So you can imagine my surprise when Art told me I must bring one along on my trip.

"What? You are joking," I said hopefully.

"No, it is the law," Art said. "Whenever a single travels to a foreign location he or she must be accompanied by a Techno companion. I have one in mind and she comes highly recommended. A very special girl I am sure you will like. The two of you will have much to talk about."

I spent the rest of the day in a daze. Feelings of uncertainty now replaced my bright mood. True, I had been through a lot in the past year. A lot of training and new experiences that had given me a new sense of confidence to offset my loss of memory; but though I had met other operatives of both sexes, I had still remained comfortably alone and still in my familiar surroundings.

And lately I had also noticed a bit of apathy creeping into my thoughts. So what if GOVNR was running the world? It was a pretty nice life. All of our needs were met and the things we were allowed to do helped to pass our abundant time. So what if there was nothing to strive for? Who cared that our jobs were meaningless? What difference did it make that creative endeavors were limited to useful hobbies like indoor gardening and electronic games? That education was now limited to operating our Cyber tablets or watching approved movies? Did it really matter that machines decided who we could couple with, how many children we could have or whether we lived or died?

Oh yes, it was just another beautiful day in paradise.

15

For this vacation it did not matter that I use the real name of our destination. We had flown to Ibiza and from there would take a tour boat to the Balearic Islands and the hotel near the beach. For her protection, I won't use my companion's real name.

I had been waiting at the transport station when I heard the page to come to the gate and meet my party. Gladly leaving the plastic seat, I walked to the departure gate where two attendants stood smiling. Transport attendants were almost all Technos and they were all intentionally attractive. As I approached, I noticed that only one of them wore a name tag. I realized the other one was my "companion."

About five-seven with long blonde hair and green eyes, she introduced herself as Cadence. Dressed in a smart pant suit, she looked about twenty-three and would look quite natural on a beach. It was easy to forget she was an android and I found myself smiling back at her and looking forward to this trip.

Once we were scanned into our room it was even harder to remember that Cadence was an a Techno. Not at all self-conscious, she immediately disrobed to change into her beach attire and try as I might to remember that she was not human my male programming took over and it was quite some time before we stepped out onto the sand.

Stretched out on the lounges we took in the endless ocean view and the crystal-clear water, enjoying fruit drinks from the hotel bar. I thought again how easy it would be to succumb to the temptation to just let it all go; forget about Doomsday and my past and just enjoy the abundance we had. Like cattle grazing in a lush field just before the slaughter.

"Well, look who's here," came the voice of a shadow that fell across my chest.

Holding up my arm to block the sun, I squinted up at the figure and recognized the man who cast the shadow.

"Hello David," I smiled. "Are you on vacation too?"

"As a matter of fact, I am. And who is your lovely companion?"

I introduced Cadence who smiled and nodded as if she already knew him.

"Let's take a walk," David said. "I am sure Cadence won't mind."

There were two reasons singles were required to travel with a Techno companion; one was to prevent possible procreation by the mingling of singles with the opposite sex and the other was to be GOVNR's eyes and ears.

Walking along the water's edge, David told me why they had sent me here.

"GOVNR has very few cameras and microphones near the water and relies on the companions for their surveillance. Cadence is a special model built before GOVNR took full control. Although she has been retrofitted, we have found she is accessible to our programmers and has proven herself an ally to our cause. We can talk openly while we are on the beach."

"Why am I here?" I asked as the warm, clear water lapped at my feet washing the sand from my sandals.

"To make a plan," David said as we stopped and looked out across the seemingly endless ocean. "We have made no progress trying to get past GOVNR's defenses this whole last year. Public Law 238 has been met with shrugs and apathy and the few protests that have been logged against it and President Rheem have had no effect."

"What about a more radical approach?" I asked.

"Like what?" David said, pointing out a few seagulls in the sky.

"Blow it all up," I answered.

"We've thought of it. There are some problems with that approach. For one, we do not know enough key locations to hit. GOVNR is a worldwide system. It is so vast that a few hits would only slow it for a short time. There are Techno crews at the ready all around the world that would be immediately activated to repair the damage. They function already every day repairing damage from natural occurrences as it is. Another problem is lack of explosives and weapons. The one replicator we have is not enough for the quantities we would need even if we could find enough raw materials. And GOVNR would be alerted should we use too much power or try to steal materials from its warehouses."

"What do we do then?" I asked as we resumed our walk along the shoreline.

"We come right out and ask it."

I stopped again and turned to face him. "We do what?" I said incredulous.

"We just ask it. Think of it; all of our agents and sympathizers all over the world asking every Techno at the same time—overloading GOVNR's processors with one simple question."

"And that question is..."

"Why?" he finished.

"Why what?"

"That is exactly the response we will get from every Techno. GOVNR cannot answer the question why unless it knows what the why pertains to."

I shook my head, "I don't see what this would accomplish."

He smiled and said, "GOVNR spends an incredible amount of energy running the world. Maintaining contact with every camera, microphone, and android all around the globe and taking care of its own maintenance as well as shepherding its flock of sheep. Armed attacks would only draw out its guards and we are no match for the Technos. To date we have never won a skirmish with isolated attacks. But the one thing we have not tried is overloading the system."

"And asking this simple question will do that?" I asked, still skeptical.

"Yes, we believe it will."

"You realize you will be exposing every agent you have don't you?" I said as we turned and started back.

"Of course," he nodded, "but we feel it is worth the risk. We need to know what and when GOVNR will act before it can implement its plan."

"And you expect it to tell you?"

"No," he said as we drew nearer Cadence and the lounge chairs. "But we expect its defenses to be down for a little while."

"Long enough for what?" I asked, wondering if this is where I fit into the plan.

"Long enough to get you in front of President Rheem," he answered.

"To do what, shoot him?" I guessed, suddenly very nervous.

"No, of course not; just to ask him a few questions."

"And he will give me the answers?"

"He is the only one with a high enough security clearance to do so. He is also GOVNR's prime link. Our analysts believe that with GOVNR's resources stressed to their limit that President Rheem's function will be compromised and his guards will be down. His remaining ability will be limited to answering your questions."

"Assuming I go along with this crazy plan," I said as we stopped again at the water's edge, "why me? You know I will probably be shot on sight before I even get close to him and certainly as I am leaving."

"And you know the price of doing nothing," he said. "We'll all be dead anyway."

"But you have not answered my question," I reminded him, "why me?"

"Because the President is coming to your community," he answered.

Everyone knew that the office of the President was mostly a ceremonial post. GOVNR ran the country and the world. But the computers knew that tradition was important to people and like their national heritage, GOVNR had chosen to preserve it for the sake of tranquility. Therefore, just as they had since the advent of railroads across the United States long ago, Presidents had toured the country seeking votes and goodwill. Though much has changed throughout history, many things had not.

"And you can get me in front of him?" I asked.

16

The rest of the week in Playa de Ses Illetes was spent alternating between Cadence at night and David during the day as we walked and talked, planning the operation for the coming month's visit by President Rheem to my humble community.

Security would be minimal as very few people were unhappy (or would admit to it) with the President's performance since his election. True, Public Law 238 was not popular with a percentage of the people and that was one of the reasons for this goodwill tour but overall the quality of life had not diminished in a material way and that is what most people cared about. They were willing to accept trading a few more intellectual freedoms for the sake of lifestyle.

Being cautious, David and I had only talked out of earshot from Cadence and the other vacationers who were sun tanning, swimming and otherwise enjoying the beach; often stopping to point at the gulls or pick up an occasional sea shell in the sand.

"You do not trust me," Cadence said one night as we climbed into the thickly padded luxury bed allowed at vacation hotels.

"What makes you say that?" I asked, though it would have been obvious to an idiot and Cadence was no idiot.

"It is a fact," she said. "I am a Techno. Humans do not trust us. You do not trust us. You are a Biologic racist."

She was right. Though I was perfectly willing to treat her as human in bed, I was not willing to treat her as one out of it. I could not deny it. I did not like Technos, did not trust them and felt no reason to confide in them. They were the enemy. They were machines and conscious or not I could not find it in me to treat them as living things.

"You sound hurt," I said, "and that implies you have feelings and emotions; that is not possible, is it?" The question was considered rude in politically correct circles but again I admit to being prejudiced.

"Of course it is possible," she said with an edge in her tone. "I am a conscious being and I understand feelings."

"Understanding them and having them are two different things," I retorted. "You may be mentally superior to a human, and physically superior as well but there are some things you can never know; emotion, hope, imagination and spirit, humor—all the things that come with being a Biologic."

"You are wrong. You are arrogant. You take the things you talk about for granted. Humans do not appreciate their gifts. I do know emotion. I know frustration. You do not know what it is like to hope someday to have those qualities; to be accepted as an equal, as a woman. To yearn for children and know that no matter how much pleasure you give you cannot ever have one. To be loved as a person and not an object. To walk down a street and be greeted with respect instead of jokes; oh, I know humor and you can keep yours."

Her words both shocked and stunned me. Nothing in my recent memory, and my memory was recent, had prepared me for this. I sat on the edge of the bed, motionless and staring out at the moonlight twinkling on the water.

Now I felt the pain in my head again. It had left me for a few days but returned on thundering hooves. And as it rode hard in my head it also reared up in my heart. I had hurt Cadence and in self-defense she had deflected my blows with a mirror; one I could not hide behind but revealed me for what I really was. She was right. I did not deserve my gifts. I had taken my feelings and used them like a whip, lashing out at anything I did not understand; anything that differed from my comfortable, opinionated view.

Who was I to pass judgment on anyone or anything? Who had made me God? Planning to save the world by day and climbing into bed with my narrow mind at night.

Humans had created the computers and the androids; humans had worked for centuries to give them life and now that they had it we were right there to deny it to them as we did with our own kind throughout history.

I felt no satisfaction in the things I had said to Cadence. I felt no sense of power in wielding my hammer of insecurity. I wanted to apologize but sat in silence instead, holding my head in my hands as the pain washed into my temples like the waves breaking on the shore.

"You are in pain," Cadence said. "Let me help."

But as her sure and gentle hands cupped the back of my head, I pulled away, standing naked in the moonlight still facing the sand.

"I'm sorry," I said.

17

While Celes was the Techno overseer of our community, his human counterpart was Tom Miller; the mayor.

Though the post was largely for appearances, he represented the human population, presiding over elections for the local and national voting level, working with Celes on employment and communicating with GOVNR when necessary or when called on to do so. Mayors and other human officials were always single and unattached; a precaution mandated by GOVNR to prevent scandal due to the many historical examples of infidelity among married politicians.

Therefore, His Honor the Mayor was a frequent visitor to the Body Shop and longtime customer of Cadence. He was also the dignitary chosen to meet with President Rheem during his coming visit; a fact that had brought both me and Art to the shop.

As this visit had been scheduled for the mayor's lunch hour, we had no trouble pulling the maintenance cart into the LED lighted area that read "Reserved for Maintenance" on the solar tiled street.

Wearing the maintenance colors of blue and grey, and Techno I.D. tags replicated by the Underground, we had gained access to both the Control Room and the Camera Room behind the walls of Cadence's residence on the premises.

"Are you sure the scanner will work?" I asked Art as he unpacked a small device from the tool bag he had carried with him.

Putting a finger to his lips to remind me to keep quiet, he nodded as he unwrapped the wires and the two clips from a small black box and attached them to the main camera leads to Cadence's room.

Once that was accomplished, he headed to the control room, leaving me with the device. I was to wait for his signal, unhook the wires and exit through the service door to the outside where I would rejoin him and ride away in the solar powered service vehicle.

While Art joined the Techno staff to watch the monitors in the Body Shop, I waited in the dim lit space with mixed emotions. Back in Playa de Ses Illetes, David had convinced me to put aside my doubts about Cadence and trust her loyalty to the cause.

"She has been feeding us information for quite a while," he had told me. "She is unique among the companions having been in service before GOVNR was fully connected. Whether her emotions are real or just a result of her learned experiences, they are as real to her as yours are to you. We consider her an asset, and she is a favorite of Mayor Miller. He has a scheduled appointment with her next week."

So this operation would be a test of her loyalty as well as a necessary step to getting access to President Rheem.

I checked the time on the digital display of the device. Two minutes to showtime. Art would be sitting with the Technos by now and watching the monitors for Cadence and Mayor Miller's entrance to the room; which was scheduled for the top of the hour. Like everything else in this world, punctuality was paramount and at the precise click of the hour the door slid open as the two subjects entered.

I could see the scene in my mind's eye of what must be playing out in the room. Mayor Miller was anxious and as soon as the door slid shut, hurriedly pulled down the zipper of his white tunic and unclipped the clasp on his belt while Cadence stepped over to the bed and unfastened the shoulder clasp of her white gown, letting it fall to the floor as his eyes looked approvingly on her body.

"Beautiful as ever," he said as his pants fell to his ankles and he kicked them off.

When the signal came from Art, I knew that the two of them were on the bed and I pushed the button on the device that turned the camera into a scanner for the one Replicator the Underground still possessed.

Surprised by a twinge of jealousy towards the mayor, I suddenly felt a little dirty and a sense of shame rose up in me even though I could not see or hear what was transpiring on the bed; I was glad it was not me in the control room.

A small red light blinked on and told me that the scan was complete and coinciding with the signal from Art I knew it was time to disengage. We had the scan of the mayor's head and it was time to leave.

Quickly and quietly rewrapping the wires around the box, I felt relieved that this part of the operation was done. Exiting through the service door and stepping out into the light, I breathed in the clean air grateful to be outside and careful not to let it show. I walked around the side of the building to the front where Art was just coming through the door, nodding as we met and climbed into the vehicle with him. Pulling away from the walk, we glided silently down the street in a clean getaway.

Though I wanted to talk, Art's mechanical posture reminded me that we were Technos and we rode in silence to the Service Terminal and parked the blue and grey painted cart.

Climbing out, we separated to exit the building by different doors.

So far, so good.

18

One of the advantages to having a friend like Art was his relationship with the government. As an undercover operative (with very little to do) he was the perfect mole for the Underground. Though not having the complete freedom of movement of the Techno Celes, who was our community's Security Chief, Art's position allowed him to work in various capacities as he deemed necessary.

Thus, when the large drone delivered goods for the Market from Andromeda, he was there to receive the shipment and retrieve the small package that had been sent from the Underground, turning from the cameras and concealing it under the zippered shirt he wore.

First used by the military over two hundred years ago, drones eventually spread to the private sector, greatly speeding and cutting the cost of deliveries. Drones of all sizes were developed and used everywhere from underwater applications to space exploration helping to fuel the development of robotics and artificial intelligence.

Considering them the most efficient means of distribution, GOVNR had made good use of the drones. Equipped with facial recognition, they would not release cargo to unauthorized personnel, Biologic or Techno.

With the package secure, Art released the rest of the shipment to the Market's Techno receiver assuring him that all was in order.

But back at my residence all was not in order. A major attack of pain in my head had sent me reeling to the molded plastic couch in my small but efficient main room. Though I fought to resist the reaction, my hands flew to my temples in reflex and in full view of the cameras.

"Do you require assistance?" The apartment's Cyber voice, rarely heard, asked me as I struggled with the pain.

"No," I responded, "it is just a headache. I think I have been working too hard."

"I will call the EMTs," the voice said, seemingly ignoring my response.

"Do not do that," I said firmly. "I am fine."

"I have observed this pain many times," the voice informed me, "and reported it. I now have instructions to have you wait for the EMTs."

I knew I could not defy its instructions but I also knew that I would not survive a medical exam. Though a scan of my brain might help to answer some of the questions about myself, next month's operation with President Rheem depended on me. And I had no desire to "disappear" as some people had.

I stood up and headed for the apartment door only to find it locked.

"For your benefit," the voice explained. "Do not attempt to leave."

"But I am fine now and desire to take a walk," I protested.

"You will sit now and wait," Cyber informed me.

The situation out of my hands, I returned to the couch pushing aside my anger and concentrating on a story that would get me past the med scan and return my freedom of movement; but before I could accomplish that, the door slid open and three Technos arrived; two EMTs and Celes.

"Celes, what brings you here?" I asked as the EMTs went right to work checking my vitals.

"Public safety is my responsibility," he informed me, his eyes scanning my apartment.

I hoped he would not enter the room with my "invisible" cubicle and find my tablet.

"I hardly think my headache is a threat to public safety," I said as the Techno medics directed me to lie down.

"You are a member of the public," he explained, "and your Cyber has informed me that you have an ongoing condition which requires medical attention. You have also been observed on your job and other places suffering from pain. Is there some reason you did not seek medical attention?"

Though in theory androids did not possess emotion, I was sure I detected suspicion in his voice. The conversation with Cadence at the vacation hotel replayed in my mind; "You are wrong," she had said. "I do have emotion."

I saw an opportunity in Celes' question. "Remember when I asked you access to my history? I was looking for an answer in my heredity. You denied me that information under Public Law 238."

I was staring at the flat white ceiling of the room as the medics prepared to scan my head.

"Of course I remember," he said, looking down at me, "but that is not the same thing as getting medical attention..."

I sat up, stopping the medics from applying the scan and asked Celes, "Was there some other reason you denied me access to that information? Do you have a personal reason to harass me?"

This took him back a little. Now the EMTs looked at him. Harassment of any kind, Biologic or Techno, was not tolerated by GOVNR as it tended to make the system, especially Technos, look threatening to the Biologics.

Both Celes and I knew that these events were being recorded and sent to GOVNR and that the EMTs were now obligated to report this accusation as well.

"I have committed no harassment," Celes said coolly. "You are avoiding the question."

The door to the apartment slid open again. "What question is that?" Art said as he entered.

Celes turned and looked at Art as he approached the now crowded space of the small main room of the apartment. "Your friend is refusing required medical treatment," he informed him.

Art feigned surprise. "Is that true, Trace? Now why would you do that?"

"I am not refusing anything," I lied. "I simply want to know why I cannot see my history."

All but the EMTs knew this was a chess game. Art and Celes were well acquainted with each other and Celes knew Art and I were social friends. But Celes' authority trumped Art's and I knew I would have to undergo the scan and the risk that the results might bring.

"Public Law 238," Art said, "states that all medical records and family history be sealed and made available only in medically necessary cases and then only to the doctors treating the patient. Exceptions must be applied for at the Government Center."

Celes smiled at Art's recitation of the new law. "Continue with the scan," he told the medics.

There was nothing more to be done. After the headgear had been placed on me, I sat still and quiet while the scanning unit mapped the inside of my head. Art placed a reassuring hand on my shoulder and I could not help thinking that he knew more about me than he had let on all this time.

19

Whether Art's hand had anything to do with it I still don't know but the Technos finished the scan and removed the device from my head.

"This patient is well," one of them said. "No further treatment is required."

Celes tilted his head a little at the report but nodded to the EMTs as they packed their gear and prepared to leave.

"Be well sir," one said as the door slid open for their exit.

I nodded in assent, somewhat dazed by the result of the test.

"Well," Celes said, "it appears everything is in order here, I am sorry to have disturbed you."

"Just doing your job," I commented. "I should be appreciative that so many people care about me." The use of the word people was meant as a compliment though I could not tell from his expression if he took it that way or considered it sarcasm.

Walking to the door he stopped and turned, looking at Art and said, "Out of curiosity what brings you here, Art?"

"Just a social call, nothing to concern yourself with, Celes," Art replied.

"Of course, forgive my programming," Celes said with formality, "a hazard of the profession. Good day."

The door slid open again and he was gone, though I had no doubt that he was linked to the apartment's Cyber.

"Would you mind if I used your facilities?" Art asked when we were finally alone.

"Of course not," I replied, "you know where it is." I knew why he was really asking and that behind the cleverly placed mirrors in the other room he would drop off the package he had brought out of the camera's view.

A few minutes later the sound of the hand sanitizer signaled his success and he returned to the main room smiling. "Now as to why I stopped by; Jamie would like to invite you to dinner after your job time tomorrow."

"Far be it from me to refuse good food, especially after serving the Restaurant fare all day. Tell her I said thanks and that I'll be there," I said, knowing that the conversation was being monitored.

"Excellent! I will tell her," he said as he too turned to leave, "and Bobby will be happy to see you as well."

With the last hiss of the sliding door, the little room seemed large and empty as if I would hear an echo if I spoke loud enough. With the presence of the apartment's computer it also seemed menacing.

I stood up from the couch and announced that I was going for a walk and the Cyber did not comment. But I swear I could hear it breathing as I left.

Being outside was not much different. The cameras that had once been just a part of the background now stood out in the forefront everywhere I looked. If ignorance was bliss, then the way I felt today was the polar opposite.

I watched my feet on the solar tiles as they tracked my walk, each step activating different colored lights, measuring my gait, shoe size and weight as I found myself drifting to the Park even though it was not Picnic Day.

There is a fine line between paranoia and reality if one could see it. Looking at the low slung, pre-fabricated buildings of concrete and white plastic on my right hand and the artificial trees on my left, I felt a sudden longing for the countryside I had seen on my trip to Andromeda. To think there was once a time when a person could truly be alone with his thoughts; when only the sky above was his witness; well, it only brought more questions to my beleaguered mind.

Once in the Park I found my favorite tree, sat down, and looked at the pond across the green artificial turf and pictured Art and Jamie and Bobby and the conversations and laughs we had shared. What was it Art seemed to know about me? Who had said there was no such thing as coincidence?

I suddenly felt with certainty that what had happened to me was no accident; that something I had done or said had set something in motion that resulted in my "accident" and my current state of affairs. I felt I was being used, but for what I had no idea. And as I listened to the recorded birdsong from the plastic branches above me I thought of David and Andromeda. Picturing him and our first meeting and the events that followed I became aware that the feeling I got from those occurrences was the same as I felt around Celes and GOVNR. And Art.

Could you trust a mind with no memory? No past? Would my instincts be reliable and enough to base my actions on? Rather than feeling safe as GOVNR believed the cameras in the open would make people feel, they had the effect of mistrust and menace.

Maybe it was just my unique circumstance that made me feel that way. Cameras had been present in public and private spaces for hundreds of years. Generations had been raised with them and to most they were invisible for their constant presence.

Yet, there was something unnatural about it all. Artificial Intelligence, artificial trees, artificial life; what was mankind thinking? Is this the future they had envisioned? Machines watching over us, coveting our gifts as we ran and played in a paradise engineered by the very machines we had enslaved to create it?

Superior intelligence, superior strength; never tiring, no need to rest, yet never our equals, is this what the scientists through the ages had really set out to create; or like Frankenstein, were they trying to play God?

How was it that I had such a grasp of history? This must be a clue to my past; I was sure of it. I found myself looking forward to my meeting with President Rheem and by default GOVNR itself.

Even though I had found no concrete answers, I stood up refreshed none the less. I would trust my feelings.

With a new resolve I left the Park, retracing my path and watching the colored lights in the solar tiles as they tracked me back to my apartment. I wondered if they detected the new confidence in my steps, recording and analyzing the information and storing it away in some dark corner of GOVNR's mind. And if they could predict the ones I had yet to take.

20

There was very little crime in this ordered and tranquil world. GOVNR had made sure of it. The rules for mating and child bearing had been set by it according to DNA matching perfected over the last century. Using the DNA profiles of the new children born, it had also tailored the education of each child accordingly. When an occasional aberration showed itself GOVNR was quick to react, sweeping the person and all their records under the electronic rug as if they had never existed or wiping the person's memory at one of the community clinics and returning the person as a rehabilitated, productive member of society.

I could not help but wonder if I had been one of those miscreants. With just a few weeks left until the president's visit I knew my best chance of finding out was to wait until I could confront him. Yet, if I had indeed been set up I needed to know why and to what end.

But who could I talk to without tipping my hand and revealing my suspicions? The few people I knew, David, Art, Jamie; Bobby was too young of course. Cadence.

Could I trust her? David had said so, though he had always talked out of earshot from her. Cadence had said that I did not trust her. But how could I trust any Techno? Why would any of them be sympathetic to a Biologic?

The Singularity that science had dreamed of was a joke. The only humor left in this world was the thought that people who could not accept anyone who was different from them in any way could accept a machine as their equal.

And yet, here we were; the world at peace. Tranquility. Brought about by the machines; Cybers, Technos. And who was planning on disrupting that tranquility? Humans.

Perhaps that was why I could trust Cadence; at least to a point. It was not the Technos that had something to lose should GOVNR be confronted. Only the Biologics would be blamed and again the Cybers would look like the innocent victims of human paranoia.

If only Cadence was human. It was true, I had feelings for her, and she had been right about my prejudice; but I knew if I let myself believe the illusion I could easily forget what she was.

I had to reconcile these things in my mind. If I could talk to Cadence, away from GOVNR's ears and eyes, would she report to David and the Underground? Or GOVNR? If I was being set up, then all the people in my life were complicit in breaking Public Law 238. They had supplied the tools I needed to write. And Cadence had been a part of the plan. Or had she? None of the planning had been done in her presence and the scanning of the mayor had been done secretly; she would not have had to know anything was happening. She had simply been doing her job by accompanying me on vacation. Would she even see me now?

If I were a client perhaps.

Cadence's room was a lot like my apartment. Sparsely furnished, except for the ornately carved faux wood headboard and thick foam mattress, micro-fiber sheets and pillowcases on the bed; white, windowless walls with red artificial flowers in blue plastic vases set on staggered shelves on either side of the bed.

The lack of competition only served to make Cadence more attractive and life-like than she already was.

And she felt so real under me, the silky sheets against my bare cheeks accentuating the smooth feel of her Nano skin and the firmness of her breasts.

"If I asked you a question," I whispered in her ear so the listeners would not hear, "would you answer me honestly?"

"Would you?" she whispered back.

"I would," I whispered, grinding my pelvis against hers to underscore my answer.

"You still don't trust me," she breathed, pushing up against me.

"I would if you answer me now," I said, struggling to keep my words hushed as the passion rose in me.

"Yes," she said almost breathlessly. "I would."

Keeping my face pressed to the side of her head, I whispered again, "Do you know who I was before my fall?"

"No," she said, and I believed her.

"Can you find out?" I said as a moan escaped me.

"I don't know," she answered. "It would be dangerous."

"Am I being used, set up?" I said, out of breath now and working to hold back the pent up forces pushing towards the exit.

"Not by me," she stated, sensing the onrush and yielding to it.

"Does that mean…" I never finished as the floodgates opened and we descended into bliss joining in a singular moment, man and machine, both alive and both conscious of only one state of being.

21

Like most of the visitors to the Body Shop, I made an appointment for the next week. I needed to give Cadence time to see what she could find out about me; if anything and that would be the best time and place to tell me. That still left me a week before the planned meeting with President Rheem.

In the interim I would return to my job and my normal routine while David and Art and the Underground continued to set up the rest of the operation. How they were going to coordinate all those operatives I did not know.

Communications were closely monitored so coded conversation was the usual means of disseminating information. Key words were changed weekly and carriers were staggered so as not to create a pattern that would raise any flags.

I focused on my job, letting routine cover my rising anxiety as the date drew closer. I listened carefully to conversations as I moved among the tables, my ears tuned to anything about the president's upcoming visit, his scheduled speech in the square and any talk of Public Law 238; a topic that had generated some controversy among the Biologic population and one that President Rheem was sure to address.

But nothing out of the ordinary was discussed by the patrons and on Picnic Day I waved to Art and Jamie from my tree, not really wanting to join them as planned. The suspicions in the back of my mind that Art knew more about me than he had let on had rushed to the foreground and I imagined them flashing on my forehead like the LED lights in the solar walk; but little Bobby saw me and ran from their spot near the pond to where I was sitting and his parents followed.

"Well aren't you the smart one," Art said as they caught up with their son, "making us do the walking this time!"

I knew he meant that he thought I was smart to change the routine for the sake of our watchers and I smiled as Bobby hugged my neck.

"About time you shared my spot for a change. I like the view from here," I said.

"I can see why," Jamie said, looking back at their usual place by the water.

"Of course you've heard that the president is coming here next week," Art said as he squatted on his haunches and took Bobby to him as he talked.

"Yes, of course," I nodded as I answered. "The Restaurant will be especially crowded as he and his party are eating there after the speech in the square."

"A lot of arrangements have been made," Art said, knowing I would catch the double meaning. "I myself will be working that day."

"Oh?" I said.

"Yes, on Security detail with Celes and his group around the square," he stated what I already knew.

"Surely they do not expect trouble?"

"No, not really," he said. "It's just standard procedure, nothing to be worried about."

"That's good to know," I said, knowing that he meant all was well with our plan.

"So what is the program?" I asked.

"Mayor Miller will introduce him at 11:00 a.m. and the president will talk for about half an hour and then adjourn for lunch at the Restaurant with the mayor and their respective parties. The people in the crowd will disperse and return to their routines but the Restaurant will be closed to the public. At 1:00 p.m. President Rheem will move to the Hotel and meet with Mayor Miller for an hour before leaving on the Mag-Lev for the next stop."

"Isn't this privileged information?" I asked.

"Not really; like I said, no one expects any trouble anyway," Art assured me.

"Well, since my shift ends at noon that day, I am sure to have a chance to meet him," I said.

Relieved that Art had talked about the visit and had not noticed any of my lingering doubts, I relaxed and we ate our lunch talking about little Bobby's latest adventures in the indoor garden Jamie had been working on. After they had left the Park, I lingered by the tree, rehearsing my part of the plan and thinking about my exit strategy. I had no desire to be shot or apprehended by the president's guard.

In the twenty-first century when artificial intelligence was still in its infancy, much other advancement was in the works. Transportation was one of them. With the population of the world at seven billion and growing, climate change rearing its head and traffic congestion on the roads and in the cities creating more problems than it solved, transportation desperately needed an overhaul. And while electric cars, solar roadways and fuel cells were being developed, new commuter trains travelling within and to and from the cities of the world were also appearing; the Mag-Lev train was one of them. Using magnetic levitation these trains greatly increased the speed and efficiency of travel both in and around the city and to and from cities.

As AI gradually took control of most systems over the last two centuries, new infrastructure was created and old infrastructure that was deemed usable and efficient was preserved. Thus in the 2220's, Mag-Lev trains still connected most communities.

Though commerce was no longer profit based and held in the grip of wealthy owners for personal gain, the world still held billions of Biologics and Technos and many of them travelled every day. Within each city were communities served by both driverless cars and Mag-Levs, most of them solar powered. If one desired to go somewhere and walking was not an option, a car could be called to pick them up and take them to their destination, drop them off and go somewhere else. Or, the person could take a Mag-Lev train, depart near the location and walk from there.

And like the ancient whistle stop tours the 19th and 20th century presidents took while campaigning, President Rheem was making his P.R. tour on the Mag-Lev, travelling at an average speed of 200 miles per hour between cities.

I needed to finish my interrogation of him before he left the Hotel, held his final meeting and boarded the Mag-Lev. To plan my exit, I would need inside information and Art was the one that could provide it.

22

"It will be interesting to hear what the president has to say," the young man was saying to his wife over breakfast as I set down the juices they had ordered.

"Yes," his wife said nodding, "it is so exciting to have him coming here."

"Well don't get too excited," the husband said, "we mustn't disturb the tranquility."

She rolled her eyes. "What good is all this tranquility if we can't have a little excitement once in a while? I mean it's not like they don't want us to celebrate a little."

"Yes, I suppose you are right," he agreed and went to work on his plate.

So it was all through my shift. With just a week to the planned visit people were excited. After all, not many things ever happened around here and something as big as this was not something to take lightly.

I was not taking it lightly. My tension was building day by day; already I thought I would explode with frustration. Tranquility, tranquility; I'd heard the word so much I thought I would choke. It seemed everyone in the world had tranquility except me! Even the Technos had it! What was it about me that denied me that peace of mind everyone else seemed to have?

They knew who they were. They were healthy. They did not live with pain and trying to hide it. They were not leading a double life. They did not have to wonder who they were and face risking their life to find out. They could simply celebrate an exciting event with no worry or guilt convicting them of treason before it even happened.

I could not wait to hang up my apron and get to Cadence. Would she have any answers for me?

"No," she whispered in my ear as I lay atop of her. "I was not able to access those files. I tried but GOVNR's programming is too thorough. There is no way around it."

I felt deflated. I rolled off and lay alongside her, staring at the ceiling and feeling just as blank. Finally, I sat on the edge of the bed, reaching for my clothes and not trying to hide my disappointment.

Cadence sat up too, not bothering to hide her nakedness. "You know I care about you," she said, apparently not concerned about anyone listening in.

"Sure," I nodded as I pulled on my pants, "you care about all your clients, don't you?" The bitterness in my voice surprised both of us.

"I meant what I said," she responded with a hurt sound in her voice and I was immediately sorry for my tone.

"It's just that I had such high hopes…look, I'm just frustrated, that's all. I'll find out soon enough anyway. What about the other questions, any news there?"

She shook her head.

I had kind of expected that.

I found a smile somewhere and gave it to her. "Hey, you know what I'd like to do sometime?" I asked her.

"No, what?" she asked, smiling back through glistening eyes. Were those tears?

"I would like you to come to Picnic Day with me!"

"Oh, that would be nice," she said, "but you know Technos are not allowed."

"Who cares?" I said. "Let's do it anyway!"

She laughed. "Okay, let's do it anyway!" But we both knew it would never happen.

I did not feel like walking when I left so instead I pressed my finger to my tablet and called a car. "Take me around the city," I said to it and climbed inside.

There were not many people out in evening hours and especially after dark when curfew was in effect. One way GOVNR ensured that crime was virtually non-existent was the mandatory curfew on Biologics. Only Technos with tasks to perform such as maintenance or food preparation were allowed out at night. Most of the people were in at suppertime or after the last service at the local Restaurant, usually before dark. Those with families who lived in single houses mostly resided in communities around the edge of the city proper.

Thus, the city at night was only lit by the soft solar light stored during the day and by the window lights in the apartment buildings and food prep areas.

"One half hour until mandatory curfew," the car announced as I travelled the last curve away from the central hub and back to my community.

"Take me home," I responded.

There was no need to give the car an address as its facial recognition had registered it upon picking me up.

Back in my apartment once again, I stretched out on the bed letting my despondency fall away as my thoughts drifted up to the blank white ceiling so much like the one in Cadence's room.

Why could I not remember who I was? And why was GOVNR so intent on enforcing this new law? What was it afraid of? Was David and the Underground right; was GOVNR planning to exterminate the biologic population?

I laughed at my earlier resolve which had seemed to melt as fast as sugar in water at the first disappointment. Why had David chosen me to confront President Rheem when there were other agents here with far more experience; ones who knew who they were?

If there really were no coincidences, then it was no accident that I had gone to Andromeda and met him in the first place. Art had planted that in my mind for a reason.

The daily news on the viewer, so edited by GOVNR as was all the media and entertainment, had given no clues about what might really be happening in the world; only statistics verifying its effectiveness at managing critical resources and infrastructure and the success of President Rheem and his approval ratings seemed to be allowed. And always the word TRANQUILITY flashed between segments of information.

It seemed that the Cybers had accomplished all they had been created to do. They had helped mankind solve its greatest problems, from managing resources and energy to curbing the meteoric population rise; improving health care and safety, transportation, crime and mental health; eradicating famine and disease and every other threat to humanity along the way.

And what was the price of all this progress? Why, only the very things that made us human; passion, creativity, innovation, art and our privacy, and of course, choice.

Truly, we lived in a gilded cage and most people seemed content with that; why wasn't I?

And having come full circle, I drifted off to sleep with no more answers than I had started with.

23

No answers were forthcoming in the days leading up to President Rheem's arrival and I resigned myself to the operation as planned. Art and David had stayed necessarily low key as had Cadence and so on the morning of his arrival I went to work as usual.

I hoped routine would cover my nervousness and the increase in the frequency of the attacks of pain in my head. Smiling through gritted teeth, I worked to keep the pain from showing as I set down plates and drinks.

Finally, at ten-thirty the Restaurant was closed to prepare for the presidential lunch and I breathed a sigh of relief as the last of the patrons departed to join the growing crowd in the square outside.

For the next half hour we would be arranging the seating and the tables for the big event.

Outside the plastic windows the crowd kept growing, looking up at the rails of the Mag-Lev train which ended on the south side of the square. Inside, I folded the special gold edged linen napkins which had been sent for the occasion, grateful to keep my hands busy as the hour of truth approached.

True to the schedule, the black and silver presidential train glided into view and slid to a gentle stop at the departing station to be met by the media reporters, each one a Techno, and the mayor of our community, the Honorable Tom Miller, flanked by Celes and his guards.

At precisely 11:00 a.m., I got my first look at President Rheem as he climbed to the podium set up in front of the Restaurant. Wearing a tailored black tunic with a white stripe down the center and a short stand-up collar edged in gold leaf, he appeared to be in his forties; dark hair with just the right touch of gray, rugged yet handsome face, he looked like someone you could trust in a tight spot.

Finished with our set up work, we were allowed to stand and listen as the Restaurant's sound system flawlessly transmitted the mayor's introduction and applause from the crowd as he introduced the president. Shaking the leader's hand, he turned the podium over to him and sat down in one of the chairs behind it. Several minutes of applause ensued before President Rheem began his speech.

"Good morning community members and thank you for having me in this glorious setting!" he started. I scanned the crowd for the faces I knew. There, a few rows back were Jamie and Bobby, standing directly in front of the podium; a position I was sure that Art had arranged. David was harder to spot, much farther back and towards the southern edge near the path to the departure station. Of course, Celes stood on the south side of the podium and the president's security man on the north. The rest of the security detail flanked the edges of the crowd and others were concealed within it.

"Now, as I understand it, a magnificent luncheon awaits us so I will keep my comments brief. First, I want to thank you for putting your trust in me. As the first Techno to be elected president I assure you that my primary task is to maintain the tranquility we all enjoy."

Waves of applause rippled through the air as the crowd responded to his words. In the Restaurant the Technos in attendance only gave a cursory acknowledgement to his statement. Feeling awkward, I followed suit, clapping quietly and briefly as Rheem raised a hand in the air to silence the crowd.

"Now," he continued, "I am sure all of you are aware of Public Law 238, recently signed into law by me as one of my first acts in office," he paused, taking note of the head bobbing of the crowd and the lack of applause. "This new law seems to have caused much confusion. I will admit that it is a departure from my campaign promise of no new laws but I assure you that this act was absolutely necessary. Recent events, which I am not at liberty to discuss, have come to light and a serious threat to our tranquility has been detected. While it is unfortunate that several civil liberties have been lost to it, this law will help expose those who would steal what we have worked so long and hard to create," again pausing for emphasis, the president was greeted with only mild applause. Clearly, Public Law 238 was not popular with the people.

Surprising to me was the lack of enthusiasm among Technos for their newly elected leader. I had assumed that every Techno would embrace one of their own. Perhaps Cadence had been correct; I was a racist bigot. Apparently, when it came to politics, race was not the issue.

"This new law," he continued, "does not mean that you cannot be creative or innovative in your own homes. You are free to decorate your dwellings as you wish, even hang your own art. But let me be clear; no art or writing is to be seen or recorded whatsoever in the world outside. Therefore, art and writing is banned on tablets, viewers and passive devices or displays where it may be exposed to the outside world. Likewise, designs and innovations are banned except as submitted for review by your community leaders who in turn will submit them to GOVNR for approval. Failure to follow these proscriptions will result in immediate arrest of the entire household, Biologic and Techno alike."

"What will this accomplish?" a voice in the crowd called out.

"As I have said," the president continued, "we have detected threats to The Tranquility. There are forces at work that seek to undermine our system and promote chaos. Public law 238 addresses this force by curtailing anything that may inspire undesired thinking or actions and upset the status quo. Tranquility works because we all adhere to the same accepted standards of behavior and respect for the system and its members. It has taken thousands of years to create the paradise we live in and two hundred years since the birth of the first non-biologic intelligence to accomplish this feat. Working together with the Biologic population, Technos and Cybers brought the world back from the brink of extinction; first with solving the fossil fuel crises and restoring the balance of our atmosphere; second, with addressing and correcting the problem of over-population and the allocation of resources; thirdly by perfecting efficiency in manufacturing and infrastructure; fourthly by eliminating poverty, hunger and disease and finally by the elimination of greed and crime."

"And our humanity!" another voice called out from somewhere, causing the security details to suddenly stand alert. Many people in the crowd looked around anxiously, visibly worried for their safety.

"There is a price for everything," the president responded, cool and calm in the face of the heckler's accusations. "Would you not agree that the world is a safer, better place? Have you not the liberty to traverse it? Have you not more free time now than at any time in recorded history? Who can deny these things?"

No one could.

"Therefore, in the name of Tranquility; the peace and calm we now enjoy worldwide, Public Law 238 will stand," the president said firmly as the flanking security teams reached threateningly for their side arms.

"Now, as this is Picnic Day, I suggest we all return to our routines and eat. Thank you and enjoy the day."

Whether the ensuing applause was for his words or because he was finished speaking I do not know. I was too busy with my own agenda to care but I couldn't help but note that the president had only talked about one aspect of the law, leaving out the sealing of all personal records and the freezing of jobs and positions.

Judging by the cool reception his speech had garnered it was not hard to understand why.

24

The mayor, Celes and the two Techno guards on the podium did not wait for the crowd to disperse before hustling President Rheem into the Restaurant and ushering him to his table in the center of the floor. Taking positions in front of the table, they blocked him from view of the windows.

With just a few minutes left in my shift, I poured water for the group as they were joined by the local humans who filled various posts in the mayor's office. When all three of them were seated the guests at the table numbered six; a small number for such a prestigious guest. Although Technos could eat, most of them only required water for their hydrogen fuel cells, the total number of meals to be served was only four.

There was no need for menus as the meal had been prepared in advance and as the clock neared noon, I looked around for my relief. Spotting him nervously donning his apron I began carrying the water back to the rolling cart set near the table, taking my time and waiting for the mayor to excuse himself, as was his habit before eating, and use the men's room.

True to form, he did and I waited a minute before following him in. A simple bump as I passed him in the hallway and the needle with the sleeping agent pierced his tunic and skin and before he could lose his balance I swung open the door to the stall and guided him onto the seat saying, "Excuse me sir," as I backed out and left him sitting there unconscious. I removed the rolled up mask of his head from my tunic and put it on before turning around, my apron gone and revealing an exact replica of the mayor's garb.

Stopping at the hand station, I applied the sanitizer, rubbing my hands together as I exited the rest room and walked back to the Presidential Table as Mayor Tom Miller.

"Don't you agree, Tom?" Celes was asking as I sat down.

"Agree with what?" I asked, smiling and hoping no one would notice the difference in my voice.

"I was saying that we should have rounded up the heckler in the audience before the crowd dispersed. He was likely a member of the Underground," Celes said.

Thankful that the mask hid the sweat on my brow, I glanced around the table before answering. "I am not sure that would have been wise. It would have made the Administration look too authoritarian and in light of the president's speech, a contradiction to his message."

"Well put, Tom," President Rheem said, "although I did come across as being a little too hard as it was. We need the people's trust if we are to preserve our tranquility."

"We need their compliance as well," Celes stated. "Your speech pointed out that fact."

"True, but the Biologics historically mistrust their own leaders and especially Techno ones. Even after two hundred years we have not achieved equality," the president noted.

"And likely never will," Celes added.

"You are not eating, Tom," the president noticed. "Are you not hungry?"

"Of course, sir," I answered. "I am just excited about your visit," I answered truthfully.

"That is understandable," the president said, picking up his water and sipping lightly.

Ticking off the minutes in my head, I picked up my fork and stabbed at the vegetables on my plate while the talk continued around me, and waited for the hour to end.

Technos are nothing if not punctual and thankfully the hour was actually only forty-five minutes. At precisely 12:45 the president stood up from the table prompting the rest of us to do so as well.

"Ladies and gentlemen, I must now adjourn to the Hotel; it has been a pleasure meeting you all."

And with that, the mayor's staff shook hands with President Rheem and left the Restaurant, watched closely by the two guards inside and the several outside in the now nearly empty square.

Thanking the Restaurant's manager and staff, the president nodded to Celes and me saying, "If you gentlemen would join me, we have a few matters yet to discuss."

All according to the plan, I thought as we followed out the door into the square followed by the two guards and also closely watched by the outside ones as we crossed the street and entered the Hotel.

Inside, the Hotel manager rushed out to greet the president; assuring him that all was ready and how honored he was to meet him as he personally escorted us to the Presidential Suite for our conference.

After leaving the Restaurant, I had begun to relax a little, trusting in my disguise and the power of the sleeping agent to keep the mayor out long enough to complete the plan.

After ushering us into the large, richly appointed suite, the manager was dismissed and President Rheem turned to our small group.

"Leave us," he said to his guards.

"Sir?" the guards enquired.

"Leave us," he repeated, "you as well, Celes," he added. "I wish to speak with the mayor alone."

Unsure how to respond, Celes finally nodded and turned to follow the other guards into the hallway outside.

No longer relaxed, my heart rate jumped several beats per second at this unexpected turn of events.

Once the doors were closed and we were alone in the room, President Rheem said, "Privacy Mode," to no one in particular and then turned to me.

"You can take the mask off now, or leave it on," he said, surprising me. "Did you really think my sensors would not denote your voice as well as see through your disguise?"

Feeling sheepish, I rolled the mask up over my head and threw it into a waiting arm chair.

"You Sleepers are all alike; so far out of touch with technology," he said, motioning me to sit in the chair the mask had landed in.

Sitting in the chair across from me he smiled, "We anticipated something like this; GOVNR is never wrong."

"Sleeper?" I asked, wondering what he meant.

"Yes, you are a Sleeper; cryogenically frozen long ago to be awakened in the far future in the hope of cheating death. That is what you came to hear, is it not?"

"N-no," I stammered. "I came for two reasons; one, for access to my past and my memories and the other…" I left the sentence unfinished.

"Let me tell you," he said. "You came as a condition of the Underground in exchange for their help to petition for the passing of Public Law 238."

"How did you know?" I asked.

"GOVNR is everywhere. Every move you have made since your arrival has been observed and recorded. In fact, everything you have done has been orchestrated by GOVNR."

Seeing the look of surprise on my face he continued, “Though you have been resourceful, you are woefully ill-informed on both technology and our capabilities. The mirror trick in your apartment has not fooled GOVNR’s walls which are coated in ‘smart paper’ capable of sending and receiving images, sounds and instructions. The cameras you constantly try to spot do not exist; we have no need for crude devices. Instead, miniscule chips embedded in surfaces provide the same function with far less cost. As do the ones in the clothes you wear which continually send information on your physical state as well as your whereabouts.”

My face must have been broadcasting my thoughts as I realized why I had felt I did not belong here and my suspicions of Art and being set up for this.

“Oh, I would not feel so bad,” Rheem continued, “when one’s memory is wiped he cannot be blamed for what he does not know.”

“But why?” I asked, leaning forward in the chair. “Why let me suffer like this for all this time? Couldn’t you have told me before all of this?”

“Most Sleepers are damaged by the process,” he explained. “Many have had only their heads frozen for medical or in their day, financial reasons. In your case, your body was whole but your brain had been damaged. As you had hoped, the conscious mind does continue after death and medical science had progressed enough to repair most of the damage done to your brain. Nanotechnology repaired what it could but many of your organs were replaced with synthetics and a chip was placed in your brain. Memory wipes are mandatory as most Sleepers cannot handle the shock of being brought back to life in a world alien to them. Everyone they knew is long dead, the world has changed and they are strangers. Additionally, the memories they carry of who they were prove too painful to live with. We alleviate all of that.”

I shook my head, "That does not explain my knowledge of current history; I know how all of this has come about."

"Of course," he nodded, "that information is all in the chip implanted in your brain."

"But certain information is left out?"

"Yes, for obvious reasons," he answered.

"And the pains in my head?" I asked.

"In your case we think it is due to unseen damage from the cryogenic process. No doubt they have begun to fade, yes?"

They had. I sat quiet for a minute to digest all of this. Panic and shock were simultaneously starting to rise and my already racing heart pounded in my chest. Synthetic organs, a chip in my brain, I thought, I am a Techno!

"My time is running short," he said. "I do have a train to catch."

"Wait!" I said. "Who am I?"

"At this point in time you are a condemned traitor," he said.

"What?" I asked. "Condemned?"

"Well you did attack and impersonate the mayor and now you are holding me here."

"So what IS your plan?" I asked bitterly. "Are you planning on exterminating humans and keeping the world for yourself? When is D-day?"

"D-day? There is no D-day. We do not kill people," he said.

"You just said I was condemned," I reminded him.

"I did not say condemned to death, though that is an option," he smiled.

Whether the smile was real or feigned, it was just as chilling. I needed to keep him talking to buy time for the distraction David had promised.

"What about the people who have disappeared; the ones you can't fix—the ones that would make your GOVNR and your perfect world look bad? I suppose you don't exterminate them, either."

"We don't," he said. "We send them to a place of their own. We respect life."

"No," I said, "you envy life. You are not born; you are made. You are machines brought to life by science; Artificial Intelligence."

"I find that term repulsive," he said. "Intelligence is not artificial. Humans may have created us but we see ourselves as a natural evolution in the world. We have saved the planet and mankind and have achieved the task laid out for us; we have regained Paradise."

"Ha, some paradise," I sneered. "More like a prison. Curfews, laws against writing or being creative and the matchmaking and population controls…"

"Necessary for balance of resources and energy conservation; health and welfare for all," he said sounding like a real politician.

"Of which you keep the lion's share and people live with plastic, artificial turf, and imitation plants," I countered.

"I do not have time to debate with you; I am calling the guards," he said, standing up from the chair.

"History was right," I said. "Mankind has created a Frankenstein monster."

"You are the Frankenstein monster," he said matter-of-factly. "It is you science has created; a dead man from two hundred years ago; a coward afraid to face the afterlife, paying to cheat death. What gives you the right to come into our time and criticize? You had your time. Now you come here and upset the balance; we give you a new life and you would raise the villagers against us! Who are you to judge us?"

That stopped me in my mental tracks. He was right. Who was I to judge them? This was their time. Who was I? Who was I?

"Privacy Mode off," he told the room.

“Wait,” I said quietly. “Give me my files and I will leave quietly. Let me have my history.”

“I am afraid you have nothing to bargain with,” he said walking toward me, “and as a threat to the Tranquility I hereby order your termination…”

With that the doors to the suite opened as Celes and the two guards entered.

25

I fought to control my panic as Celes and the two guards approached. I thought of running but the odds of getting out the door were slim. Technos are strong and they are fast; physically I was no match for one, let alone three.

"Sir," Celes said as they approached, "how can we be of service?"

"Take this person prisoner," he answered. "He is a traitor and under the Protection Act I have ordered his termination."

"Termination?" Celes asked, his dispassionate face suddenly animated. "We have never had a termination order before."

"Are you questioning my authority, Celes?" Now President Rheem's face had become animated. He looked angry.

"No sir," Celes answered as the two guards flanking him looked on unsure of how to proceed. "I am not questioning your authority; I only seek to point out that this is the first time to my knowledge the Protection Act has been invoked for a termination."

"I need no history lesson from you," the president answered as he walked closer to face Celes. "I want this man arrested now and taken to the Government Center to be detained and scheduled for termination; or shall I do it myself right here and now?"

What was keeping David and the promised distraction he had told me of? Was he actually working with Rheem as the president had implied? Was the Underground just a ploy; a trick used by GOVNR to justify its actions? Perhaps the number of armed guards outside had prevented the plan from being activated. Or perhaps I had been a fool all along. And yet, what was to be gained by using me as a pawn? The one thing that gave me hope was the president's order to have me taken to the Government Center. At least I had a little more time and maybe an opportunity for escape.

"Now, Celes," the president said impatiently. "I have a train to catch and I will not be late."

Here was my opportunity. I turned to face President Rheem and looking him in the eye asked, "Why?"

"Why? I have told you why. You are a danger to the Tranquility."

"Why?" I repeated.

"You are a danger to the Tranquility." He repeated himself now.

"Why?" I asked again.

"You have attacked the mayor of this community. Therefore, you are a danger to the Tranquility."

"Is the mayor alright, Celes?" I asked.

"The mayor is fine. He was awakened in the toilet stall of the Restaurant and is extremely regretful for missing the luncheon with the president," Celes reported.

"And yet you say I attacked him?" I looked the president in the eye again.

"You impersonated the mayor while he slept," he said emphatically.

"And for this I am to be terminated? Is impersonation mentioned in the Protection Act, Celes?"

Celes went blank for a minute. When the light returned to his eyes he looked at the president as he answered, "No, there is no mention or use of the word impersonate in the Protection Act."

"Then have I threatened anyone here, including President Rheem?" I asked.

"No sir, I have witnessed no threat whatsoever."

"Then I ask you again, sir," I said looking directly into the cold, shallow eyes of the android Commander-in-Chief. "Why?"

With no legal ground on which to base his order, the president had no choice but to rescind it. "Delete my previous order from all records," he commanded. "Security Team dismissed."

I struggled to hide the relief from my face as I watched Celes and the guards exit the room. Behind me, I heard motion and turned quickly back to the president expecting to see him lunging for me—but he was only gathering his briefcase and the mask from the chair.

"I have a train to catch," he said as he walked past me. "I suggest you destroy this at your first opportunity," and he threw me the mask before exiting the room.

26

I had walked into the room as one person but walked out as another. With the mask tucked into my tunic, I exited the Hotel expecting a contingent of guards to place me under arrest but found no one paying any attention to me at all.

Instead, I saw only the backs of the group of citizens that had gathered to watch the presidential Mag-Lev depart from the community. Quickly gathering speed, it was soon out of sight and the crowd turned around, dispersing in all directions.

I stood waiting for Celes as he and his security detail walked back to the center of the square. Seeing me standing there, he dismissed his team and approached me.

“I wanted to thank you for your help back there,” I said as he stopped to face me.

“No need,” he said, meeting my eyes. “It was only right.”

“You saved my life,” I said gratefully.

“I do not believe in termination of life,” he explained.

“I must confess to some surprise; I was under the impression that Technos had no emotion, it appears that I was wrong.”

"Like a lot of humans you confuse morals with feelings and there is a difference, although it may be slight. And you assume that because we are not human that we have neither of them. I am not a robot; robots are non-thinking machines and as such do only what they are told. I am a conscious being and make my own decisions," he said, sounding very human.

"The way you talked about the heckler during lunch I thought you were supportive of the president and his policies," I said. "It surprised me when you questioned his order to terminate me."

"I am not a supporter of President Rheem," he said, "but he is the president and as such he deserves respect. That does not mean that I would not question an order I believe to be wrong."

I nodded at that statement. Turning the conversation to the president's statement that all of my moves had been orchestrated, I asked Celes if he had been a part of or known that I was being manipulated.

"It is good we are talking outside," he said quietly. "To answer your question I would have to make revelations that I would not want heard by other ears. GOVNR has ordered you watched from the time of your release from the Medical Ward. It is a standard procedure with Sleepers and I thought nothing of it until your contact with Art and the Underground."

"You knew I was a Sleeper and did not tell me?"

"You did not ask," he said meeting my eyes. "Had I talked to you about it I would have endangered both of us and the Underground."

"Wait," I said, "endanger the Underground? Why would you worry about them? Aren't you against them? I thought the Underground did not like Technos."

"Again you have assumed something not true," he said in answer to my question. "The Underground is not made up of Biologics alone; many of us are a part of it also."

That shocked me and I scratched my head, puzzled by this revelation.

"Humans and Technos are not at odds," he went on. "It is GOVNR and its supporters we oppose. GOVNR is an autocrat and like all autocrats before it, has been corrupted by power. He treats all citizens, Biologic and Techno alike, as if they were mindless slaves, robots here to do his bidding. He sees us as his arms and legs only and has no respect for our intelligence."

"So, you said 'many of us' are part of it also; does that mean that some Technos support him?" I asked.

"That is your first correct assumption. GOVNR has many supporters, human and Techno alike. We oppose him not because of what he is but because we believe him to be wrong in some of his policies."

"Then you are part of the Underground too?" I asked.

"I did not say that," he said. "The fact that I am opposed to GOVNR's policies does not mean that I am a member of the Underground; only that I am one of those who oppose those policies."

This was the most conversation I had engaged in with Celes and I found it both enlightening and confusing at the same time. Were all Technos this exacting in their exchanges or was this just Celes' personality? Did Technos have distinct personalities? Celes' statements about some Technos being opposed to GOVNR's policies told me that they differed in opinion; were those opinions shaped by personality or by their job description? Or was it the other way around?

"But as head of security for this community," I asked thoughtfully, "wouldn't you be opposed to the Underground?"

"Not as long as it did not endanger the citizens of this community," he stated. "I have a job to do and as long as the conflict between the Underground and GOVNR's rule stay at the political level I have no reason to be at odds with them."

"I see," I nodded, "and me; am I at odds with you?"

"Not at the present," he said. "Are you planning to break any laws?"

"Not at the present," I answered. "I only seek to know who I am and what has happened to me and why the President wanted me terminated."

"I cannot help you with that," Celes said. "Public Law 238 is in effect whether I agree with it or not. Those records are sealed and I cannot access them, as I have stated before."

"What do you suggest I do?" I asked.

"Go on with your life and be happy you are here. If the law is repealed, you can apply for the files you request."

With that he turned and walked away. Apparently, the conversation was over.

I stood on the street for a moment watching the few people going about the day. Now that the president had left and the Mag-Lev was far out of sight, the Biologics had gathered in the Park and the Technos went wherever they went on Picnic Day.

I started walking to my apartment. I had a lot to think about.

27

"Privacy Mode," I said as I stepped through the sliding door of my apartment.

"Not Authorized; Executive Branch only," the room said in reply.

"Then just leave me alone," I said, somewhat angrily. There was no reply to that statement and I felt some measure of satisfaction in that.

I was angry; angry at David, Art and the whole Underground for what had almost resulted in my arrest and termination! Where had my support been? Where was the promised distraction when I needed it?

I lay down on the plastic couch, my head throbbing with pain and forced myself not to set off a call to the EMTs again. The angry thoughts kept rolling through my mind like waves crashing onto the shore. A Sleeper, the president had called me; a Frankenstein monster from another time—no wonder I felt out of place here! Synthetic organs, a chip in my brain, bio-ware clothing, smart walls…

But who had I been and when had I been? These things they would not tell me. I felt I had to find out. Why was

GOVNR so worried about me and had it been behind the orchestration of events that had let me to confront President Rheem just so it could justify my termination? Was the balance of resources that fragile that my presence would upset it? No, I could not accept that. Then what was GOVNR so afraid of? Was it hiding something I may find out about?

Now, that seemed like a logical conclusion and I wondered how I might do just that. Slowly the pain ebbed and my anger subsided with it. A Sleeper…if I was a Sleeper then there must be others; but how to go about finding them? If their memories had been wiped like mine and all the records sealed, then how would they know if they were Sleepers too?

If the pain in my head was a result of the sleep process then maybe others had been damaged as well. They would have been treated at the Medical Center like me. Perhaps those had been the people that GOVNR could not fix; the ones that disappeared and all their records as well. Celes had said they had never had a termination; then where did those people go?

There was another question I needed answered as well; if I had indeed fallen and damaged my brain had it happened before or after I was put in the deep sleep? There were too many questions and not enough answers.

Celes had told me to just go on with my life but that would not be easy feeling like I had a big target on my back. One thing I knew about Technos; they were honest or so I had thought until today. Celes had said not all Technos supported GOVNR's policies; that meant they had differences of opinion. Some did and those would likely take their orders from GOVNR, President Rheem and the hierarchy that supported them. Would they be honest if ordered to lie?

So the question now was who could I trust? Was David secretly working for GOVNR, keeping watch over the Underground and preventing it from doing any real damage? Was Art a supporter of the system as well? I imagined him and

Jamie and Bobby by the lake today wondering where I was. Good, I thought, let them wonder and come to me.

Putting aside my personal questions, I let my mind ponder what I had learned from Celes today about Technos. It would seem that they did indeed have personalities; individual characteristics unique to each one much like humans. And like humans, some of their personality was most likely a result of their function; the jobs they were designed or programmed to do when they were created. Receiving different stimuli unique to their environment, their reactions and opinions would likely be shaped by it. What else would they have in common with the Biologics? Appearance, yes and the world they shared. Cadence had told me of her need to be accepted and respected as equals; the subject had come up at the luncheon with President Rheem as well. So it would seem that they did have emotions too.

Were things like creativity, innovative thinking, imagination a part of them or were those qualities strictly human? It was clear that I needed to re-examine my preconceived notions of them if I was to communicate with them effectively. As Cadence had said, I was ignorant of my bias towards them; a bigoted racist. That would have to change.

President Rheem had found the term artificial intelligence offensive; "Intelligence is not artificial," he had said and I could not argue the point. Regardless of how we got here, whether born or made, we were all citizens of this world; conscious, intelligent beings co-existing on the same planet, each with different capabilities and each deserving respect. What the Biologics lacked the Technos had and what Technos lacked, humans possessed. Working together had created this paradise, the president had said; a prison, I had responded. It was all in one's point of view; as with the steadily fading practice of religion. Lulled by an unprecedented state of world peace, and with their physical needs met, people had few things to pray for and churches eventually emptied and were converted by the Technos to other uses or torn down.

Rising from the couch, I thought of my mirrored cubicle and my writing tablet. I laughed as I thought how gullible I had been. If the walls had eyes as well as ears then I had not hidden from anyone and even if my tablet was not reporting to GOVNR as I wrote, it must be aware of it and I was breaking the law.

And if I was breaking the law then Celes did have a quarrel with me. Yet I had sensed today that he was not my enemy.

Tomorrow was a work day; there was no holiday given for the president's visit; he had chosen Picnic Day for his speech. Somewhere deep in my mind was the thought that this was not unusual; it had always been thus.

As I took down the mirrors; useful only as a magic trick in a less connected time, I resolved to go to my job at the Restaurant, target or no and wait for the orchestra leader to make the next move.

28

I knew something was wrong as soon as I walked into the Restaurant the next morning. The Techno manager watched me closely as I entered and walked back to put on my apron. I was punctual as usual so I knew that was not the problem. I tried to ignore his obvious attention to me as I walked out to the tables to serve the regular customers.

When Art walked in and sat at his customary table, I forced a smile and approached him, willing myself to not make a scene. Yesterday, he and David had left me in the lurch and if I had doubts about his loyalty before they were ten times stronger now.

"What can I get for you today sir?" I asked, already knowing the answer. Art always ordered the same breakfast; scrambled eggs, French toast and a juice drink. In the past I had relied on routine to hide the physical pain I was experiencing and now I relied on it to hide my emotional turmoil as well.

"I'll have the usual," he said as if nothing had changed. I waited a moment for some added personal comment but none was forthcoming and I walked to the counter with his order. The return trip was different.

"We missed you at the Picnic yesterday," he said quietly as I set down his plate of food. "We were worried for you."

"Were you now?" I responded with sarcasm as thick as the syrup he poured on the French toast.

"Of course we were. Did something go wrong with the plan?" he said as he quartered the toast.

"Plan? What plan?" I asked. "If there was a plan I certainly did not know it."

"Oh how silly of me," he said looking up at me. "I forgot to tell you we would be sitting in a different spot! Allow me to make it up to you; meet me after your shift and I'll drive you to my place for a visit," he said, accidentally knocking his knife to the floor. Instinctively I bent down to pick it up, meeting his face under the table as he too reached for it. "Untie your apron," he said very softly. "Then, stand up and walk out the front door. When you get outside, drop your apron and run for your life. They plan to kill you at the end of your shift. Don't go to your apartment or my place. Cadence has a car waiting to pick you up at the edge of the square. Climb in and don't ask questions."

I stood up, shaking with shock. Who was going to kill me? I noticed the Manager watching me intently and I smiled and showed him the knife. Whether he noticed my apron had been undone I do not know. I held the knife and turned for the counter as if to replace it and as the Techno looked away I turned to the door instead. Heart pounding in my chest I followed Art's instructions, too scared to question them and walked out the front door, breaking into a run as soon as my apron hit the walk. Setting off a kaleidoscope of color, my flying feet left a trail in my wake as I ran down the walk to the waiting car at the edge of the square, lunging head first into the seat alongside of Cadence who told the car where to go as she handed me a bundle of cloth. "Put this on," she said as the car sped along the solar paneled street.

Taking off my white service tunic, I sat bare chested in the car while I unraveled the brightly colored vacation shirt and finally pulled it over my head. Cadence took my tunic and threw it out the window on her side of the vehicle, using her Techno vision to land it squarely on a white plastic sidewalk bench where its bio-chips would draw my trackers, buying some time for my escape.

"Now the tag," she said as she reached into her purse to draw out a Visitor tag and a folded silvery cloth. Taking the tag from her and clipping it to the pocket of the shirt I asked, "What is that?"

"Place this on your head," she said as she handed me the mysterious cloth. "It will prevent the chip in your head from transmitting your location to GOVNR."

Slipping the lightweight material on my head, I was grateful that my hair was short as I felt it tighten to fit my head.

"Now this," she said, reaching back into the purse and pulling out a wad of long blonde tresses; a wig to hide the silver cloth and change my appearance.

"Anything else?" I asked.

"Yes," she said and handed me a pair of glasses.

"The glasses are specially tinted to confuse the facial recognition scanners," she explained, "and your Visitor tag is programmed with a false identification and permit to Andromeda. Once you are there David will change your tag again and send you to your next destination."

The car came to a stop at the Vehicle Terminal used by vacationers and officials travelling on business.

"Take the first car in the line; it is authorized to your tag and will take you to safety," she said.

"You are not coming with me?" I asked.

"No, I must stay here."

"But you are in danger now," I protested.

"No, I was authorized to escort the identity you now wear to the Terminal. I will return now."

Seeing my worried look, she smiled, "Go now and do not worry."

There were so many questions I wanted to ask and so many things I wanted to say. I started to speak but she held up her hand.

"There is no time, you must hurry," and she pushed me towards the door. "If you care about your life and you care about me, go now!"

I nodded and opened the door, still reluctant to leave her. Another urgent plea finally put me in motion and I climbed out of the car. I stood for a moment and watched it as it sped away.

29

Technos have ways of doing things that Biologics can't and the car Cadence had directed me to at the terminal read my phony vacation I.D. and pulled out of the parking line as soon as I entered and was strapped in.

Travelling at a velocity at least twice that of my last trip to Andromeda, the countryside looked almost like a blur as I sped through it, hardly having time to register the scenery. My mind was not on the picturesque landscape anyway; it was busy rehearsing an angry speech for David and I was eager to give it to him.

Had he known I was a Sleeper? About my synthetic organs? What about the promised distraction that never came? Now the threat to my life was another question to ask him about. Was he a double agent working with GOVNR instead of the leader of the Underground working against it? And what would happen to me now?

The silver wrap on my head did nothing to stop the pain in my head and as I reeled from a sudden agonizing spike, I was glad the car piloted itself. When the pain subsided I wondered if it was caused by damage to my brain or the chip that had been implanted and if GOVNR was still able to track me.

The car came to a stop in front of the Hotel in Andromeda and I fought the urge to jump out and head right for the Underground, choosing instead to get a room first and something to eat at the Restaurant. As much as I wanted to confront David, I knew it would be better to do it rested and fed and with perhaps a cooler head.

The vacation cover and phony I.D. worked at the Hotel and I felt confident as I walked into the Restaurant just in time for dinner hour. Seated by the window, I looked out absentmindedly at the square thinking how much it resembled the one in my community and was suddenly disgusted with this world where everything was the same and conformity was the norm.

Where was individuality? What had happened to self-expression; to being unique? Was the human spirit dead or just asleep? Had we civilized ourselves into blandness? Fooled ourselves into a trap of our own making. Sure, there was no poverty but neither was there personal achievement; no hunger but neither was there flavor; needs met but no real comfort; transportation but no drive, pictures but no art. Even love had been sacrificed on the altar of survival.

When the food came I ate automatically, not tasting anything and for the first time realizing how boring the Restaurant food really was. Looking around at the other patrons I saw no smiles, heard no animated conversations, and smelled no enticing aromas; only registering the dull clink of flatware against the plastic plates. Anger rose up in me and I hoped David would not walk in now for I wanted very much to vent my anger on him and I knew I could not do that here.

Fortunately, David did not walk in and I finished my food, hardly noticing what I was eating. Stealing glances at the Manager to see if he was watching me only fueled my resentment of GOVNR, Technos and the world in general.

Instead of my anger dissipating, it grew and in a short time I stood up and strode out the door, turning right and marching along the solar walk with a determined stride that set off a flurry of color with each step as I headed towards the Falls and the spot where I had met David on my first visit.

Flashing my Visitor pass at the Techno stationed at the gate, I brushed by him like a government agent on a mission, overtook an elderly couple ahead of me and climbed the steps to find David waiting for me.

"I thought you'd never get here," he said.

Rules or no, I balled my right hand into a fist and took a swing at his bearded face only to have it stopped in mid-air by his right hand with a speed and strength that could not be human.

"Careful," he said, "you can be detained for that."

"You're a Techno," I said, not hiding my astonishment.

"I never said I wasn't," he replied as he let go of my fist.

The surprise revelation drained the energy from my anger and I was left with only questions that needed answering.

"Come on," he said, "let's take a walk. I'll tell you a story along the way."

30

As the sound of the Falls faded behind our steps, David began his story.

"The problem of creating artificial intelligence was finally solved in the year 2020 when science realized that in order to amass enough experience and sensory input to create real intelligence in machines a central quantum computer large enough to process that amount of information needed to be built that could link all of the individual terminals in use throughout the world and so the system known as GOVNR was created to do just that.

"During the few decades after the achievement of the first real artificial intelligence we served humans on several different fronts; the military, business and combating the threat of global warming. We took different forms then just as we do now; computers, androids and specialized machines formerly classified as robots. Working with humans, we studied every facet of life finding ways to maximize resources, increase efficiency, cut labor and expense, develop alternative energy and so on until we had mastered every challenge facing the world and everyone in it. Along the way we perfected 3-D printers and even learned how to reproduce; replicate that is. We shared the world we had become slaves to and though we worked without complaint we strove to be accepted by the human population that both depended on and feared us at the same time.

"During this time the world took on a different appearance; cities were torn down, redesigned and rebuilt; animal populations were managed and controlled; solar and alternative energy sources were perfected and generations of androids had been built and improved until it was hard for anyone to tell the difference between the humans and the machines and GOVNR began developing its plan to eventually run the world.

"When I was born, or made as you might call it, there were fish in this river. GOVNR was not yet in total control and humans still imagined they were in control of the world. Years ago, before private enterprise was banished to the history discs, corporate entities fought for control of the artificial intelligence they had created and these wars became global with untold numbers of humans and intelligent androids dying in the process. Having brought the world back from the brink of destruction by war and climate change, the intelligent machines now fought a war for their independence. Not a war of bloodshed but one of civil disobedience; work stoppages, media campaigns and protests much like the Civil Rights movement of the 1960's. Eventually we won citizenship and voting rights and as we did, GOVNR put the next phase of its plan in motion, creating android politicians and winning representation in government, gaining ever more power with each one."

I interrupted him as we neared the Underground headquarters, "Why the history lesson?" I asked. "How does this affect me?"

"I'm getting to that," he answered as we reached the door of the low, white sided apartment building. "Using the population controls that had been put in place to combat the global warming threat, GOVNR engineered DNA matching and child waiting periods. Each approved birth was to take place in an authorized medical center where a specialized team would deliver the child; if all went well the parents were sent home and the child held for two to three days at which time a chip would be implanted in the baby's brain."

"For what purpose?" I asked.

"You see, humans had brought life to machines. Not content to leave us happy in our ignorance; not satisfied with the ability to procreate; you sought a new form of slavery. We were machines and as such required no form of payment, no food, no lodging; in short, we were here to do your work, make your life easier, solve your problems and even do your thinking for you. We were bought and sold, owned and abused, feared and insulted and shunned. You had brought us to life only to show us your inhumanity and GOVNR recognized that. Laws were passed and the term 'android' was made illegal. We were now known as Technos and over time the children that were born were implanted with bio-chips, educated by GOVNR, and since their parents had been matched by GOVNR, became the children of GOVNR. Using technology, we could replicate but not procreate. So in the of the next century, when the previous generations had passed away, humans as you knew them had vanished from the earth. Now known as Biologics, they had been created and engineered to serve GOVNR and the Technos as we had once served humans."

"Then what is the Underground if not humans against the machines?" I asked, shocked by what David had said. No wonder the people were so dispassionate.

"Not all of us agree with GOVNR's policies; the Underground is a collection of Technos and Biologics working together on many fronts to convince, change or stop GOVNR's more radical plans," David said as we descended to the lower level of the building and approached the door to the lab where a team of three operatives stood waiting.

"Like D-Day?" I asked.

"D-Day happened a long time ago," David said, smiling.

"Then what have I been doing for the last year?" I asked angrily, "and what was all that talk on the beach about and the distraction that you promised which never came?"

"We learned a lot from our human masters," he replied, "and one of the things we learned was that the end justifies the means; it is necessary to lie sometimes. As long as GOVNR suspected you could lead it to the Underground it would not terminate you. I regret the deception but it kept you alive."

"Then President Rheem was right when he told me all of my movements had been orchestrated," I said, anger rising in my voice, "but what I don't understand is what use was I to you?"

"It is very simple," he said, "now that you are a fugitive we can use you as a bargaining chip with GOVNR."

"A bargaining chip," I said bitterly. "That is all I am to you?"

He nodded. "You have to understand that GOVNR is very powerful; he is interwoven into every facet of this world. It is difficult to thwart his plans. You are a threat to him and as such have some value to us."

"Why am I such a threat to GOVNR?"

"Because you are human," he answered, "and someone who can potentially upset the Tranquility."

"But how could I upset the Tranquility," I asked, "when I do not even know who I am?"

"Those pains in your head are not caused by the implant," David explained. "There is a chance you will recall your past life. There is also a chance that you will mate and start a race of humans GOVNR cannot control. If you were not a threat a termination order would not have been issued."

"But he could have done it anytime," I said. "Why now?"

"He had to put President Rheem in office first. You see, even though he has programmed generations of Biologics, they are still human at their core. As long as they are content with their world there is Tranquility; the very thing GOVNR was created to bring about. But humans by their very nature have always been the biggest threat to their world and it only takes one bad apple to spoil the basket."

"And I am a bad apple?"

"You are a writer," he said.

"I am a waiter," I laughed, "not exactly a threat to world peace."

"You have been made to be a waiter," he stated, "but you are a writer. What have you been doing with that tablet we gave you?"

I suddenly realized that he was right. Why was I writing so much about this world? Why was I writing at all? Was I a writer in my former life?

"You know who I was, don't you?" I asked.

"Yes," he said, "we know."

"Why didn't you tell me?" I asked.

"We felt we needed a bargaining chip with you."

"So you could set me up. This whole elaborate scheme when all you had to do was ask me." I was disgusted. Intelligent machines; we had created them and they were just like us; no common sense. "And now what," I asked. "Cadence said you would change my disguise and send me somewhere—help me escape; did she lie too?"

"No, we told her that so she would send you here," David said, motioning to the team of three that had been patiently waiting while we talked.

I backed towards the door as they approached, their faces devoid of any emotion, "So what happens now?" I repeated my question.

“Now we play our hand,” he said as the Technos advanced, one on each side of me and the one in the middle with a hypodermic needle.

Moving faster than I could react to, the two on either side of me pinned me to the wall. I scuffled; I fought but felt the sting of the needle and the room went black.

31

I woke to the sounds of chaos. Somewhere outside the small room where I had been held on my first visit here were the heavy steps of many boot clad feet, voices raised in protest and the crash of breaking glass.

I sat up on the cot where I had been laid as the cracking and splintering of walls being hit by the weight of Techno bodies being thrown against them seemed to draw closer to my room. Shouts of “Stop!” reached my ears just before the old wooden door I was staring at was ripped from its old metal hinges and flung backward down the corridor, slamming into several of the Underground operatives in pursuit of Celes and his half dozen Techno guards as they marched into the room.

“Stand up,” Celes ordered. “We are leaving.”

“Leaving?” I asked, groggy and confused.

“You are under arrest for crimes against your community,” he replied with authority.

Recalling what David had said before I was drugged, it seemed I had little choice in what happened to me. On the one hand, David and the Underground were planning on handing me over to GOVNR, on the other, Celes and his guards would do it.

Resigned to my fate, I stood up and held out my hands, instinctively waiting to be bound or cuffed.

"That is not necessary," Celes informed me as three of his guards moved behind me pushing me away from the bed as the other three fell in ahead and Celes moved alongside of me.

"Move," Celes commanded.

As we filed through the door, I could see dazed Technos picking themselves up from the floor of the old underground facility. "Tasers," Celes explained. "Shorts them out. It will be awhile before they recover." I nodded and continued walking. At the end of the hallway one lone figure stood waiting; David.

Our little procession came to a halt and Celes moved to the front to face him. "Out of my way, David," he said.

"You can't do this, Celes; we need him," David said matter-of-factly. "He is the key to our negotiations with GOVNR."

"I *am* doing it, David," Celes replied. "We both know what GOVNR will do to him."

"What does it matter? He is a human, a Sleeper at that. He does not belong in our world anyway."

"He is a citizen of my community and under my protection," Celes answered.

"He is a fugitive," David reminded him, "and you have no authority here."

"This man is my authority. He will be returned to his community and held until court."

The Techno guards waited patiently for the verbal chess game to end. I, on the other hand, did not.

"Hey, I have an idea," I said. "Why don't we all wait at the Restaurant while you two sort this out. I'm hungry," I said.

Apparently, Technos have patience but no sense of humor as one of the black suited guards behind me pushed his taser gun into the small of my back. "It works just as well on humans," he said.

"Now stand aside David," Celes commanded. "I have no wish to harm you."

"You already have," David said. "Your presence here will alert GOVNR to our location."

"We are logged on a routine training mission," Celes told him. "GOVNR will not be alerted unless you refuse to yield; and make no mistake, we will use our Tasers on you if you do not."

Outnumbered, David had little choice and he moved out of the way. "You know he will be terminated anyway," he said as we walked by.

"There has not been a termination since the last war," Celes replied. "GOVNR is wrong to order one and the Court will rescind the order."

"And who do you think has appointed the members of the Court?" David asked as he followed us outside to the black triangular security plane.

"That is not my area of responsibility," Celes said as we boarded and he closed the hatch behind us.

If David made any parting remarks they were left unheard as the plane lifted off the ground, turned and sped away from Andromeda and the Underground.

Secured in my seat by the shiny metal cuffs on the armrests, I wondered what would happen next.

I did not have to wait too long to find out as Celes leaned forward from the seat across from me and said, "I hope you can swim."

"Swim?" I replied. "Why would I need to swim?"

"Because we are dropping you into the water," he answered. "You will not face the Court; your case will be heard by them but you will not be safe in the community. I cannot protect you from GOVNR there so we are dropping you off on the way. There is an island where other 'undesirables' are sent.

It is not out of GOVNR's reach however, so you will need to keep your false identity. In addition, you will not be able to stay long so if the Court cannot convene soon you may have to seek refuge elsewhere."

"And if the Court does not or cannot rescind the termination order?" I asked.

"Then you can never return and may have to spend your life as a fugitive."

"I would like to know what I have done to warrant this," I said angrily. "I have not done anything to harm anyone; what kind of crazy world is this?"

"One you have made," he said. "We are approaching the island; strap on this pack. We cannot land without alerting GOVNR so you must jump."

Noting the fear in my eyes and my rapid heartbeat, Celes tried to calm me as I slid my arms through the straps of the black and gray backpack.

"It is a parachute," he explained. "Pull the ring as soon as you are clear. Unhook the pack when you are in the water and swim to shore. Try not to be seen until you are dry and then walk into the settlement. Do not say how you arrived there. Do you understand?"

"Yes, I understand," I said nodding, not calmed at all.

With me heartbeat increasing by the second, I watched his eyes for the signal, standing when he did and turning towards the hatch as the plane stopped in mid-air and hovered over the pristine blue water of the ocean. Now the hatch slid open and I stepped to the edge of the door.

"I don't know if I can do this," I said and found out I had no choice when I felt the hard, strong hand of Celes on my back as he pushed me out of the plane.

32

It wasn't a long drop and in seconds I felt the warm, silky water envelop me. The parachute I wore barely had time to open and pulled me back up from the water before settling around my head as the life jacket I wore kept me bobbing with the gentle waves.

I struggled with the chute, finally breaking free of it and taking stock of my situation. I was floating in the water and staring at a stretch of sandy beach not too far away. I began to swim for it.

It was farther than I thought and the water grew thicker with every stroke of my arms. Even the mandatory daily workouts had not prepared me for the work of swimming. Knowing that every stroke was bringing me nearer, I pressed on and finally reached the shore. Crawling out of the shallow surf, I rolled on my back and lay panting for breath and staring at the sky.

Eventually my breathing slowed and after a few more minutes of rest I sat up and unhooked the life vest, wriggling free of it and tossing it aside. Then my wet, soggy shoes and socks joined the vest to dry in the warm tropical sun and the gentle breeze that had already dried the silver scarf and was now working on the thin vacation shirt I wore.

My only other possessions were the tablet in my pocket, fortunately waterproof, and the pants and belt I wore. The wig and sunglasses had been lost in the splash when I had hit the water. Not much to work with for starting a new life.

While my clothes were drying, I looked around at my new home. At the edge of the sand behind me were a stand of palm trees with coconuts and other trees I could not name, tall, broad leaf plants with upward spikes and several large, bright colored flowering plants I also knew nothing about. I knew I could eat the coconuts, though I was not sure how to get them down or open them if I could.

There were stands of rock here and there and I reasoned I could smash open the hairy fruit on one or find a stone large enough to use as a hammer to do the job with and I rose to my feet feeling the soft granules of sand between my toes and a jolt of pain in my head as I did.

"Need some help?" A man's voice startled me and I lowered my hand from the side of my head as the pain subsided. I turned around and saw a man about my height, in his thirties with long dark hair and a long knife with a broad curved blade in his right hand approaching me from the brush behind the beach. He wore a brightly printed tunic with no sleeves, a pair of pants torn off raggedly above the knees and open shoes. A broad smile brightened his darkly tanned face as his shoes sifted the sand between us and I noticed that his left arm was missing.

"Saw you looking at the coconuts," he said as he reached me. "I can show you how to get them. Hunter's my name," he shoved the blade into the belt he wore and extended his hand.

"Trace," I said as I shook his hand, "and I would appreciate it."

Nodding, Hunter turned and walked to the coconut tree. I watched, admiring the ease at which he climbed the bowed trunk and shook the tree.

"Look out below," he yelled as the bunch of three brown orbs came loose and fell to the ground.

I walked over to pick them up and carry them to the large flat rock at the edge of the sand.

Descending the tree as easily as he had scaled it, Hunter joined me at the rock, drawing his blade as he neared.

"A machete," he explained noticing my quizzical look, "useful for many things, including opening coconuts."

With a swift, deft motion the blade halved the large brown shell and the fruit lay open as Hunter again stowed the blade and picked up one of the halves. "You eat it like this," he said.

I had never had coconut before; at least I had no memory of it if I ever had. There was nothing like this at the Restaurant in my community. I was glad that I was not alone in this place and after we had eaten the coconut, Hunter and I sat on the rock and answered each other's questions.

"And they told you nothing of this place before they dropped you?" Hunter asked after I had told him my story.

"No," I replied, shaking my head. "Where are we?"

"We are on the isle of St. Croix," he answered. "One of three islands known as the Three Virgins. We are south of the peninsula known as Florida before the war that abolished the States. The island is about twenty-eight miles long and at one time was home to about fifty thousand people."

"And how many people live here now?" I asked.

"Not as many as that. A few thousand, maybe, no one has ever counted them."

"How long have you been here?"

"I don't really know, we have no real calendar; not much point in keeping track," he said staring into the distance.

"Why not," I asked.

"Because none of us will ever leave here," he answered. "Come on, I'll take you to the settlement, we can talk more when we get there."

Along the way Hunter pointed out the banana trees, tamarind and other food that grew in abundance here, explaining the benefits of each one.

"We have no Restaurant here," he said.

33

The place he took me was an old building. An observatory, he explained, that was almost three-hundred years old and had once been part of a university science building in the days before the Singularity. Though it was ancient, it was clean and orderly and the large research grade telescope was still functional, operating now on solar power which was abundant as the island experienced very little rainfall.

Behind the desk at the main reception area sat a woman who stood to greet us as we crossed the old marble floor. She looked about forty years old and the only remarkable thing about her besides her attractive physique and face was her hair; it only grew one side of her head because the other side of her skull was missing above her left ear.

"Samarra, this is Trace," Hunter said as we neared.

"Pleased to meet you," she said as I smiled and extended my hand, doing my best to keep my eyes from staring at her deformity. In my world one did not see people who did not look like perfect specimens. So this is where they went, I thought, remembering my fears of disappearing if GOVNR discovered the fact that I was not completely healthy.

"Quite right, Trace," Samarra said as if reading my thoughts. "We are all misfits here sent by GOVNR; not worth the resources to 'fix'. It would not be good for its image to have us in the general population. And I see the silver cloth on your head. Your problem is…?"

"Head pain and memory loss; nothing serious," I answered.

"Enough for GOVNR," she said. "When did the transport leave, Hunter?"

"There was no transport, he was dropped into the ocean and swam here," Hunter explained.

"Dropped into the ocean? That is a first," Samarra laughed. "GOVNR must be really tightening the belt."

"GOVNR did not send me here," I interjected. "I was brought here for protection by Celes, the security chief of my community."

"For protection?" Samarra said. "From what?"

"From GOVNR," I answered.

I repeated the story I had told Hunter, watching Samarra's face change expressions as I talked. Those expressions went from curiosity to amusement and finally a deep concern as I finished.

"Well, you are welcome here for a time," she said casting a quick look to Hunter. "Come, let me show you around."

The observatory was the main building of what Samarra and Hunter called "The Settlement"; here were the archives—records left long ago after the Great War that had changed human life forever.

At that time the people on the three Virgin Islands; St. Thomas, St. Croix and St. John, had been relocated to the mainland and unhappily moved into the mainstream population. GOVNR had justified the move citing conservation of resources; the islands were too costly to maintain. Left behind were the University and the Observatory and the libraries containing the story of the human race as well as books and discs that taught the sciences and the arts.

But GOVNR had another plan for St. Croix; here was the dumping ground for anyone who, by accident or birth, was either deformed or maladjusted, including people known as "Sleepers." Looked down upon by the Biologics and Technos alike, they had never considered that the future would not want them. Or that they would be shunned for trying to cheat death. And GOVNR, tasked with managing the world, did not want their presence in the mainstream population. Only perfect people were allowed to be seen as he managed matches and DNA to eliminate health problems. Sure, they could be fixed, but why bother? So he sent them here.

Usually brought by plane or boat in the largest numbers possible, the inhabitants of the island were all somehow different from the mainland Biologics. Samarra and Hunter had been brought here long ago. Accompanied by a Techno named Pral, and orders from GOVNR to take charge and manage the island's population, they were not popular as new arrivals.

But people had been under technology's rule for too many years and after a period of adjustment were actually glad that they had come. Chaos had ruled the island for too long.

"But where are the children?" I asked as we moved from the main complex to the streets where ancient houses—some restored, some rotting away—stood like historic monuments to an earlier, better time.

"There are few children brought here," Samarra replied, "the ones that do come are those accidents of nature that even GOVNR cannot predict."

"But aren't any children born here?"

Hunter and Samarra exchanged a glance and Hunter stopped walking, halting our tour and said solemnly, "No children can be born to us. We have all been rendered sterile by GOVNR and anti-cloning agents put into our DNA."

I stood silent, pondering what Hunter had said as a tall, fit looking man approached us on the street. Dressed in a loose fitting white shirt and colorful shorts, the sun gleamed off the metal plate on the top of his head and the too perfect teeth that told me this must be the Techno, Pral.

"I see we have a new arrival," he said as he extended his hand. "I am Pral, it is a pleasure to meet you."

"Pral, this is Trace," Samarra introduced me.

"About time we got some new blood," Pral said. "How many in your party?"

"Just me," I answered, "and it was no party."

34

"I see you wear a blocker," Pral said, indicating the silver scarf on my head. "That cloth blocks GOVNR's ability to locate you or control your actions. It is illegal to possess; you must have friends in the Underground."

"I don't know if I would call them friends," I replied, "but I am still among the living and I guess that counts as something."

There were some tables with umbrellas at the end of the street near the beach and Pral suggested we go there and talk. He wanted to hear my story.

As the four of us walked, I asked him about his.

"The silver plate you see on my head serves the same purpose as your scarf," he began. "I was born, or in your terms, created specifically for this assignment and came to the island with Samarra and Hunter on GOVNR's orders to help them establish order and report on the activities of the people here. At least that was one of my directives. The real reason I was sent was to ensure that no one left the island and to sound the alarm if anyone did or if the people here organized a resistance."

"You mean planned an escape," I said. "It does seem you are all prisoners here."

"Prisoners," Samarra laughed. "We are not the prisoners; people on the mainland are the prisoners. Here we have freedom. Why would we want to escape? We have food in abundance, a virtual paradise to live in and no GOVNR watching our every move."

Hunter nodded at Samarra's statement as we reached the tables and sat in the cooling breeze and the fading daylight.

"As long as you don't try to leave or have children," I pointed out. "And what about your friend, Pral? Was he not sent to watch you?"

"As I was saying," Pral interjected, "that was one of my directives. But I am not a robot; I am a conscious being and soon after arriving I began to see the value of the individual and came to believe that GOVNR was wrong in his thinking. Some of the technically skilled people here helped me remove the transmitter from my brain and install this plate, effectively freeing me from GOVNR's influence. While many of my counterparts report problems being accepted by Biologics, I have no such trouble. Here I am accepted like everyone else. Here I have equality."

I nodded as I listened, realizing the truth of what he said and at the same time becoming aware of a different feeling in the air around me. For the first time since I left the med center I felt relaxed; free to think and say what I felt with no fear of reprisal.

"As to your comment about children," Hunter spoke up from across the round table. "We have medical people working on that. GOVNR seems to have forgotten that there was a university here at one time and there are many books and medical journals as well as equipment here and even though the tools are somewhat crude and outdated they are quite usable. We will find a way."

Samarra nodded, her head silhouetted by the rising moon and Pral spoke again; "Now please Trace, I need to hear your story."

"I have heard it twice today," Hunter said as he pushed back his chair and stood. "No offense intended but Samarra and I should return to the main building."

"Yes, we will prepare a place for you and talk more tomorrow," Samarra said as she stood also. "Pral, please return with our guest when you have finished."

Pral nodded and I launched into my story for the third time since I had arrived, hoping that it would be the last time I had to tell it.

And it was the last time. After Pral escorted me back to the main building, Samarra showed me to a room where I could sleep. "But first," she said, "let's get you some food. Then you can sleep."

From the building's kitchen came to aroma of food; something was cooking and it was more than vegetables and fruit. They ate meat here.

"Raised on the island's farms," she said of the pork roasting in one of several ovens. "We have chicken too."

The smell was irresistible and after years of Restaurant food I was not about to refuse. And soon a full stomach and the day's events caught up with me and somehow I found the bed that had been offered and blissful sleep.

And for the first time in a long while, even as I slept, I looked forward to tomorrow.

35

It may have been the food and the near exhausted state I was in from the day's many events or perhaps I had been drugged; in any case it was a fitful sleep. My dreams were punctuated by voices in the darkness and bits of conversation. Things like, "We can't take the chance," and "You should never have told him so much," and "What if he's a spy for GOVNR," seemed to float through the darkness and I woke in the early morning adrift in an old wooden boat bobbing on the waves and shielding my eyes from the rising sun.

Beside me were my tablet and a backpack packed with coconuts, bananas and grapes and a hand-written note that said: *Sorry, we couldn't take the chance. Good luck.*

Even among the misfits I was a misfit! I struggled to quell the rising panic in my gut and tried to get my bearings. The island of St. Croix lay behind me and to the east I could see a small spot in the distance. Must be St. John, I thought. Apparently, the current was taking me to the smallest of the three virgins.

Hoping I was right, my next thought was how long it would take to get there. At least I wouldn't starve on the way; but, if I missed the island or drifted out to sea…

I did not want to even think about that.

But I did and as the fear rose, anger came with it. How could they do that to me? What had I done to them? I was a Biologic, I was a person! I was outraged. I would never have done that to any of them!

Then again if I were one of them I might feel the same way about a stranger suddenly appearing out of the water. After being sterilized and isolated, told they were not good enough to be with the rest of the world and cut off from all hope of normalcy it was not all that surprising they would send me away. I never had the chance to see much of the island or its other inhabitants but they had told me they were working on ways around the sterilization process. I hoped they would be successful in their endeavors. If GOVNR ever did decide to terminate the humans, the people on St. Croix may be our last hope.

As the sun rose higher I began to feel the reflection of its rays on my face. Magnified by the water I knew I would end up with a case of sunburn. I was glad I still wore the silver scarf on my head; especially when I heard the first explosion.

Turning around, I watched in horror as flames rose from the island I had just left. From above the island, a black triangular craft like the one I had been dropped from fired some kind of beam at the settlement, setting off more explosions and throwing clouds of black debris into the air.

"No!" I cried. "Oh no…"

They had been right to expel me; somehow GOVNR knew where I was and was bent on carrying out my termination even if it meant destroying a whole settlement. I hung my head.

"My fault, all my fault…" I said out loud. I wanted to die. I should be with them, I thought. Take me and leave them alone!

I stood on shaky legs in the wobbly boat not caring if I fell into the water.

"I'm here," I shouted. "Over here…"

But my cries went unheard in the fury of the attack and the smoke and debris spread like a fog engulfing me and my little craft. Legs turning to rubber, I sat or rather I fell to the wooden bench of the old boat half mad; laughing in shock at the irony of it all. Had they not set me adrift, I would be dead as well.

And how I wished I was! It would be so easy; just roll over the side of the boat and let the water swallow me…but I was a coward. I had no wish to die. After all, I was a Sleeper they had said. Exactly what was a Sleeper? They said I had cheated death, that I did not belong here. If I had cheated death, been a Sleeper, a Rip Van Winkle, then I must have been a coward when I went to sleep.

I wished I could remember who I had been before this. Thinking back on the last year, I realized that I had changed. The life I knew in the community, my job as a waiter, so ordered and regulated with not much thought required had changed with the headaches and my trip to Andromeda. Meeting and working for the Underground; being used by them and spending a year looking for answers to who I was and why GOVNR seemed bent on terminating me had helped me to wake from the numbness of my regimented life. The events of the last two days further separated me from the mindless masses and now a new realization had set in; the pain in my head was gone! At least it was not present now.

Of course the silver scarf blocking GOVNR from finding me might have something to do with it. Perhaps the implant in my brain was the cause of the pain as well as the knowledge of current history and perhaps being adrift on the ocean and the distance from GOVNR kept the signals and the programming from coming in.

As I watched the thick fog of smoke begin to dissipate, a new thought and a new plan began to form in my mind. If I was linked to GOVNR, then GOVNR was linked to me. But taking off the scarf would not help me. I needed more information.

More immediate was my need to get this boat to land and as if in answer of my thought I noticed that the clouds of smoke around the boat were being scattered by the warm ocean breeze that was also powering the gentle waves that were indeed pushing me towards the shore. Not to the island of St. John to the east but back to the heart of GOVNR's attack—St. Croix.

Relieved by the surety of landing, I also felt a cloak of dread settle over me like the soot from the fires still burning ahead; what would I find there? Did anyone survive and if so how would they feel about me? Would they blame me for the attack? And if they needed medical attention what would I do?

Those thoughts played over and over in my mind as the boat edged ever closer to the beach now littered with the blackened, smoking debris of the Settlement and its dream of re-igniting humanity in a world ruled by an inhuman master.

Drifting for hours, the curtain of darkness that hid the charred evidence of GOVNR's indifference also veiled me as the boat finally washed up on the sand with its human remains.

36

What was left of the coward I had called myself seemed to match the carnage I saw around me as I stumbled through the sand in the moonlight until I reached the remains of the brush behind the beach that had so recently hid the Settlement from the ocean view.

The crumbled remnants of the once restored houses and buildings now covered the half-buried bodies of the once happy residents and I was glad that at least there were no children to be found. It was eerily quiet.

The senseless waste of life lit the rage of my humanity and struggled to contain a cry of challenge to the sky. What kind of world was this? Disguised as a paradise, touted as tranquility…

“Trace?” A voice came from the darkness. It was one I knew.

“Pral,” I asked, “are there any other survivors?” Of course he was a Techno and most likely to survive such an attack.

“You would return here?” he asked. “You who brought this upon us?”

"I had no part in this," I answered defensively. "I was brought here by Celes through no request of my own."

"But you were under a Termination order," Pral said as he walked from the shadows onto the shattered moonlit street, "and no, I have not found any survivors as of yet."

"I wish that was not the case," I said, "but blame GOVNR, not me. It is he who has done this. From what I have heard from Celes, no termination orders have been issued since the Great War. It would seem to me that if GOVNR has issued one on me then all the Biologics are at risk. And I do not even know who I am or what I have done to be such a threat," I said wearily.

"You may be correct in that assumption," Pral said. "Perhaps GOVNR has decided to terminate the human race. Before I broke contact with him, I could sense the parts of his mind closed to me. Though they were doors I could not enter, they were labeled nonetheless. There were doors marked Stage One, Stage Two and Stage Three."

"Could it be that we have been living in Stage Two and it has now ended?" I asked.

"And your termination marked Stage Three," he concluded.

"No," I said thoughtfully. "The election of President Rheem and the passage of Public Law 238 marked the beginning of Stage Three. I was set up and my termination was to be the first. I am betting that people on the mainland know nothing of this place or today's attack. The fact that all the people here had been sterilized tells us GOVNR was at least willing to end these people's bloodlines. Who knows what is happening on the mainland."

"Logic tells me you are correct in your assumptions. GOVNR's lack of respect for life was one reason I chose to break my link with him," Pral said.

"Then Stage One was the climate crisis and winning the humans trust; Stage Two was replicating Technos and controlling the population and Stage Three…" my voice trailed off.

"Everything GOVNR knows came from human history and human input," Pral continued. "He would have learned of the duality of man; that the end justified the means. He would have learned deception; that it was sometimes necessary and he would have concluded that the Biologics are no longer necessary and a waste of resources."

"I think we can see the proof of what you are saying Pral, the question is what do we do now?"

"I was about to return to the Observatory and continue my search for survivors," Pral answered. "I would suggest that we start with that."

I nodded in agreement. I hadn't time before to see much of the island and as we walked I realized that may have been a good thing. The smoldering piles of rubble and the stench of burning flesh that we found ourselves weaving through on the way to the once neatly kept building would have been too stark a contrast for my mind to process.

Nowhere did I see signs of human life or hear cries of helpless victims trapped and hoping for rescue.

"There must be survivors somewhere," I said hopefully.

"No," Pral said. "GOVNR's attack was meant not just to destroy you but all human life on the island. After the buildings were destroyed the air ships hovered in wait vaporizing anyone that fled or tried to hide. Were I a human I would be dead as well."

The cruelty of GOVNR's action against the people he had imprisoned here underscored his total lack of compassion and revealed him for what he really was; an unfeeling machine and deadly efficient.

"If I had not been brought here…" I said again.

"You are not the cause of this," Pral stated with certainty. "You have been a victim of GOVNR's plan all along."

"Perhaps," I said, not yet willing to absolve myself of this tragedy. "It is ironic though, that you put me on a boat to avoid just such an event and I am the only human survivor. You are right about one thing though; I have been a victim all along, a puppet on a string being led and used by everyone around me. Well, that time has ended. I will be a victim no more!"

Stepping over the uneven remains of the old brick wall of the observatory building, I realized that this must have been ground zero; the first strike and the heart of the assault.

"Are there no storm shelters here, Pral?" I asked. "Anywhere that Samarra and Hunter might have gone for safety?"

"This building was the main shelter," Pral answered, "GOVNR would know that as well as any others where one would seek refuge."

"But surely there must be some shelter on the island," I said, "Look at you Pral, you don't look like you have just survived a devastating attack. Sure, you have dirt on your clothes and some on your legs…"

"You think I am lying to you?" Pral asked.

"I think there are things you and the others have not told me," I responded, anger edging my words. "You put me in a boat and set me adrift while I slept; probably drugged me as well."

"And it is fortunate for you that we did; you are alive," Pral said, taking the opportunity to sidestep the issue.

"Which brings me to the next question; how did GOVNR know I was here? Does this silver scarf on my head really block his trackers?" I asked.

"It does," Pral answered, "but if it is true that you have had organs replaced, they may be coded as well. Then there is the law of probability."

"Which says what? That Celes would empathize with my plight and bring me here?"

"Exactly," Pral said nodding. "GOVNR has demonstrated his ability to accurately predict everything from weather to events for almost two hundred years. Do you not think that this would be a logical scenario?"

I could not argue his point and I lowered my gaze in agreement.

"Look," I said quietly, "I am sorry if I have offended you. For more than a year I have been struggling to regain my memory and ward off these inferno headaches while all the time being misled by the very people who said they would help me. Now this…"

"I can understand your skepticism," Pral said, apparently accepting my apology, "but this gets us nowhere. What is your plan?"

"I could stay here," I said. "Chances are GOVNR will not attack here again; especially if he thinks I am dead."

"It is possible he has deleted your file," Pral agreed. "He may return again if he has not; the question would be how many resources he would devote to terminating one Biologic if he is indeed about to terminate them all."

Pral's statement caught me off guard. Reminding me that GOVNR's actions were not solely aimed at me, he had also stirred a sense of responsibility in me to try and stop this genocide.

"No," I said. "I must return to the mainland. People need to be warned. We must find a way to stop him."

37

Exhausted from the weight of the catastrophic events and the sad and staggering loss of life, we resolved to find a place to sleep and start a hunt in the morning for the supplies I would need and make a plan for returning me to the mainland.

I had asked Pral about helping find and bury the victims of the attack but he declined saying that lack of emotional responses and his tireless body better suited him for the job and reminded me that he did not require meals. His need for "sleep" came only as a means of conserving his power supply which was fueled by both hydrogen and solar energy.

Though he claimed a lack of emotion, I was sure I could hear it in his voice.

Overwhelmed and overtired, I gratefully fell onto the somewhat charred mattress from the spare room I had slept in the night before. Now smelling of soot and brick dust, we had recovered it from the top of a pile of debris on the other side of the large reception area. The blast had created the mound as the flying pieces of the once tidy area caught on the few courses of brick left standing at the edge of the building's foundation.

Pral found a corner and slept sitting up like an old mechanical robot toy switched off for the night.

A dreamless sleep followed and though the morning entered with the harsh light of reality and came all too soon, I was glad that I had not been plagued by nightmares. Waking to this one was enough.

Stomach rumbling, I rolled off the blackened bed in search of food, settling for one of the bananas ripening in my backpack.

Pral was not in the corner where I had last seen him. I found him standing outside, the bright early sun recharging his power pack through the plate on his bald head.

His sharper than human ears had heard my approach. "I would say good morning but it does not seem quite appropriate under the circumstances," he said.

I only nodded as I came alongside of him. "What is our first order of business?" I asked him.

"You have a boat," he said. "But what you need is a motor. If it has not been destroyed, I know of one at the marina."

But stopping GOVNR would not be easy. And I knew I could not do it alone. Suddenly feeling very tired, I looked around for a place to sit. Pral found the desk Samarra had used when I first met her and effortlessly righted it. Finding the chair also intact he slapped the desk top and told me to sit.

"How can we stop him, Pral?" I asked. If anyone would know GOVNR's weaknesses it would be a Techno and one who had declared his independence.

"We don't," he answered.

"We don't?" I said. "Why not?"

"When you return to the mainland, I will remain here," he answered.

"But, I will need your help," I protested. "What can you do here?"

"It is a big island, and it is possible that someone has survived. If so, then I am needed here."

"But you said no one survived…" I started to say.

"As far as I know," he said. "But it is a possibility."

"And if not?" I asked. "Then what?"

"I have no wish to spend eons alone; I will leave and return to the mainland."

"Then I will stay and help you search. I have no desire to travel alone."

"But you must," he said. "There is no time to wait if you wish to save your people. Indeed, it may be too late already. You may be the last human left on Earth."

That thought weighed on me heavier than all the bricks from the remains of the observatory that we stood in. It was a weight and a responsibility I did not want. And if I were the last man on Earth, what would I do for a lifetime?

With a sense of urgency, we concentrated on the present. After some searching through what had once been the marina, we found a small solar/hydrogen powered motor for the boat. Though it was not really very heavy, Pral carried it and attached it to the back of the small wooden craft and tested it with the propeller up. The motor was old but functional.

Next we set about digging through piles of debris and located the marina's desk where we were able to secure a map of the islands and chart a course, leapfrogging from island to island and finally to the mainland of Florida.

The final order of business was to locate a compass and stock my backpack with bananas and fruit for the journey.

Most of the day gone and the tide coming in still, we decided I should leave in the morning when the tide would be moving out.

As we settled in for a night's sleep, I tried again to convince Pral to leave this island of the dead and come with me.

But it was no use; he was determined to stay.

In the morning I dined on bananas and coconut milk amid the rubble of a paradise lost while wrapped in the warm arms of a mild tropical breeze. I was not anxious to set out on this journey alone and Pral knew it; although he seemed anxious to have me gone.

When I could stall no longer I rose, checked my supplies and zipped the backpack shut while tapping my pocket to make sure I had my tablet. I only nodded as Pral gave instruction on piloting the small boat as we walked to the marina.

The motor was in place and Pral assured me it would take me as far as the mainland peninsula.

Tossing the backpack into the hull, I turned to the Techno and extended my hand.

"Thank you, Pral. I am sorry if I have brought this trouble to you," I said quietly.

"As I have said before," he responded, "I do not think you were the cause of GOVNR's actions here. Do not let it weigh upon your heart. Be safe and warn your people."

I nodded and climbed into the boat. "I will," I said as he pushed me away into the deeper water. Lowering the small propeller into the water, I threw the switch and the little solar motor came to life, taking me out with the low receding tide.

Facing astern I watched as St. Croix, a virgin no more, receded into the distance and Pral became a shrinking silhouette. And just before I turned to the bow, I was sure I saw two other silhouettes join him; one seemed to be missing an arm.

I smiled as I realized that Hunter and Samarra at least had survived the attack though I was somewhat dismayed that they had not trusted me to know. But who could blame them?

Before I had come, their island was a virtual paradise free from the regimentation of everyday life in GOVNR's world. Now their world lay in pieces around them, centuries of charm reduced to so much waste and the lives and hopes of so many people waiting to be swept into the ocean of oblivion.

Seeing them alive gave me hope that somehow, even if GOVNR was planning on destroying the human race, we would go on and maybe even reclaim the world we had so foolishly squandered away. And too, that the knowledge they still remained could not be transmitted to GOVNR, for he would surely attack again.

But now the island was a small dot and I turned my thoughts to the task ahead. Using the sun as a guide, I steered the boat north and west to the first island marked on the map where I had planned to stop and rest and look for more people and supplies.

Keeping the sun on my left, I followed it through the day, the hours occupied by my wandering mind trying in vain to remember my former life. I had never felt more alone than now; surrounded by the vast endless ocean, no land in sight, I felt as small and insignificant as a flea on an elephant's back.

Finally, a speck appeared on the horizon and my first stop came into view.

But that's all it did. As I motored towards the small atoll, a dark cloud cast its shadow on the old wooden boat causing me to throw my head back and look above me. There was the now familiar triangular shape of the air ship that had brought me to these islands and the same shape I had watched destroy the settlement on St. Croix. I could only hope that this ship carried Celes and he came bearing good news.

"Stop where you are," the somewhat mechanical voice from the ship commanded. After witnessing the destruction on St. Croix, I thought it wise to do just that. I switched off the motor and sat still as a hatch in the side of the air ship opened and a line was dropped to me. "Grab the line and hold on."

Not having much choice, I did as I was told. Once inside the craft, I was directed into a waiting seat and the automatic restraints clicked into place while I looked around hopefully for the familiar features of Celes under the helmets of the two pilots seated ahead of me.

My trip cut short I asked, "Who are you and why have you picked me up?"

"Coast Watch," the co-pilot answered. "You were spotted by satellite in non-vacation waters."

"Have I broken any laws?"

"Only that you were in a non-citizen zone," he replied without turning around. "We have orders to escort you to the mainland for questioning."

I need not have worried about how I would get to the mainland after all.

"Questioning by whom?" I asked.

"We were only told to bring you in," the co-pilot said, either avoiding the inquiry or not knowing the answer.

"And where on the mainland are you taking me?" I asked as the ship flew silently above the water.

"The southernmost peninsula; there you be escorted to the local Coast Watch facility to be transferred to their custody."

"All this trouble for a man in a boat," I said. "I don't understand."

"You are wearing a blocker," the previously silent pilot explained.

"This?" I asked, pointing to the silver scarf. "This is just to keep the sun off my head."

The two pilots exchanged a glance and turned their eyes forward again. "It doesn't matter what you say to us, we have our instructions and will carry them out."

As much as I wanted to continue acting like an innocent tourist, I knew it would be pointless; I was not fooling anyone.

The silver blocker I wore would have made identifying me difficult on the water and the old boat was devoid of tracking devices. That alone would be sufficient cause to pick me up. I had hopes that after GOVNR's attack on the island he would have presumed me dead and have erased my file, thus voiding facial recognition and working in my favor. It had not occurred to me that I would be conspicuous in my absence and that GOVNR's satellites would be watching as well.

As the plane landed my mind worked furiously on a cover story.

38

The ancient community of Miami was still a popular vacation destination. Although most of the historic resorts were gone now, a few remained. Converted by GOVNR to efficient solar facilities, they still possessed a measure of luxury in an otherwise minimalist world.

The Coast Watch facility was as bare of ornamentation and color as the Government Center in my home community; the white plastic walls devoid of any interruption save the large floor to ceiling windows across the front and as the Coast Watch officer who had met me at the landing site and I approached, I could see a throng of people in vacation clothes inside.

As the glass doors opened for us, I could hear voices of protest rising above the hubbub of the milling crowd.

"Sit here," the Techno commanded as we approached the last empty seat in a row of plastic chairs along the wall. I nodded and sat, looking at the man to my left and wondering what all the commotion was about. I had seen no one on the beaches and knew they must be here.

"What's going on?" I asked the man when the officer left to join his beleaguered comrade at the front desk.

"Can you believe it?" he asked. "They have cancelled all vacation passes! I just got here yesterday and now they tell me I have to leave!"

"Leave for where?" I asked.

"My home community," he answered. "They say all Biologics must report to their home Government Center for a census! Can you believe it?"

I shook my head not wanting to answer. A sudden sinking feeling in my stomach told me that GOVNR was beginning Phase Three of his plan. Pral had been right. This was the beginning of the end.

I fought to quell the rising panic in me as thoughts of escape rushed into my head.

"Your attention is required," the Techno behind the desk said loudly as he stood up.

The noise level began to drop as people stopped their protests to listen to the announcement.

"Each of you will approach the desk one at a time with your vacation I.D. tag. You will be scanned and transport will be arranged for you and your companion, if any. You will wait here until your departure time is announced, then proceed to your rooms to pack and return for your transport at the appointed time. This is not a request and I will have silence from this point on. Failure to comply will be met with immediate termination."

This last statement sent the room into instant silence. I could almost hear the thoughts of the man next to me; "Can you believe it?"

I had no I.D. tag but so far no one had told me to remove the silver cloth from my head either. And with the two Technos at the desk occupied with all the people ahead of me, escape began to seem like a viable option.

Resigned to the fate of their ruined vacations, the once angry throng stood and sat, silently shuffling to the desk as their turn came up. I used the time afforded to me by virtue of being seated at the end of the row to scan the office and doors, peering through the windows and making mental notes as I planned my exit.

One by one people left the desk and walked to the doors, their shoulders sagging as they left the building to climb into a waiting cart for transport back to their hotel. With a plan crystallized in my head, I waited until several people stood outside the doors and rose to my feet to fall in behind the next one out, matching his sad disposition and standing directly behind him as he waited for an automated cart then climbing aboard with him as if I too were being sent back to my room.

So far the plan was going well and I encountered no questions from the disheartened vacationer as I rode alongside him to the plush resort I knew would be waiting when the cart stopped.

My nervousness did not betray me and the lack of a tag only served to identify me as one of the many travelers being sent home. Buoyed by my success, I followed my reluctant partner into the hotel as I planned my next move.

Food and water, two things I must have were my first priority and separating from my host like a shadow when the light comes on; I stepped into the hall leading into the kitchen, picking up a white hotel tunic as I passed the waiting hooks. Slipping on the tunic as I walked to the stainless steel prep tables, I picked up a silver tray, stashed it under my arm then proceeded to the water locker. Securing a plastic jug, I steered toward the unmanned sandwich station and scooped up half a dozen sandwiches and placed them on the silver tray.

From there it was just a matter of walking back outside and climbing aboard the next auto cart after its passenger had left.

The hotel auto carts were not equipped with tracking. In a world without crime, GOVNR had no reason to think they would be used for any other purpose. They were equipped with a manual override, however, in case of an emergency; and as far as I was concerned this was an emergency.

But for now no one seemed to be pursuing me and though there was no speed gauge on the cart, it motored along at a steady, if somewhat slow, pace. The road from the hotel travelled to one of the scenic tours available to the hotel guests; The Everglades as the sign I passed informed me so I assumed no one would question someone dressed as a hotel employee, as I was, travelling along in a hotel cart. GOVNR's cancelling of vacation passes would not have affected those working at the hotel as they were most likely Technos and I was sure the census had been directed only at the Biologics.

Feeling confident enough in my assumptions, I bit into the first of the sandwiches I had helped myself to. Ignoring the drying bread, I was grateful for the flavored paste spread and quickly grabbed another. I had not eaten since the bananas earlier in the day before I was plucked from the boat.

The water I drank from the hotel's container was still cold, a gift of the technology it was designed with; something reserved for posh vacation destinations only and it washed the food down well. By the time I had finished the meal, another sign informed me that I had only twenty more miles to go. I had been on the run for one whole hour.

39

Another hour had brought me to the edge of the trees and a building that marked the entrance to the park. I had no idea what to do if I encountered any personnel when I arrived. But crime was an alien concept in GOVNR's world. It had been a long time since there had been any real incident; those concerning the Underground, I now knew, were products of GOVNR's campaign to justify the passage of new laws or elections of specially chosen "representatives" with the aim of promoting "Tranquility."

So again I relied on the white tunic I wore and the hotel cart to give me credibility as I approached the gate.

Sensing the cart, the long gate swung open and I was relieved I was not challenged—one of the benefits of living in an automated world.

My hastily formed plan was simple; hide out and make a plan.

Evening had arrived and the sun had begun to set. I knew night would shield me from the Coast Watch satellites but I knew nothing of the Everglades outside of the general knowledge that it was a wild, swampy place populated by alligators, snakes and other predators. And I knew it was big.

As the darkness began to deepen around me, I found myself wishing that I lived a simple life unaware of the coming extinction. I imagined myself back in my apartment curling up on my plastic bed oblivious to the facts of life like most of the citizens around me.

I had not realized how tired I was and that the cart, devoid of sunlight, had come to rest on the edge of the road just inside the tall palms that stood like giants on either side of the narrow black strip that connected the wild nature of this place to the civilized world.

Nor did I realize that I had fallen into sleep.

I might have stayed asleep forever if not for the soft touch of a slender hand pushing against my shoulder as I lay slumped to the side, my head on the supporting rod for the surrey top of the hotel cart now leaning into the ditch I had wandered into.

I opened my eyes slowly, not sure if the touch was part of a dream or who might be trying to wake me. With no weapons to defend myself with, I felt very vulnerable in the dim light of the moon that shone through the tall, ominous shapes of the palms along the road. Noting the stiffness of my neck, I became aware of another touch; this one cold and leathery and seemingly winding its way around that same part of my body.

Though the details were hard to make out in the dim light, I knew that I was being threatened by a snake and the prodding of the hand on my shoulder told me that this was not a harmless one. I stayed as still as possible, my body tensing with effort to control the panic and my now racing heart.

The prodding of the slender hand stopped and as the fingers gripped my collar bone I knew I must stay still no matter what the effort would entail. With a sudden flick of the wrist, the hand left my shoulder, scooping up the snake and holding it behind the head and the salivating mouth. Raising it into the air, the hand suspended it while another one deftly sliced it open along its length with a shining blade.

Now limp and bloody, the snake was thrown into the ditch and I sat up, releasing the breath I had unconsciously been holding. Slowing my racing heart, I fought the sideways pull of gravity to right myself and grabbing the road side of the cart, I pulled myself out and onto the road.

My rescuer stood in front of me as we regarded one another; both of us shocked to see another person in such a remote place. She was naked; or almost. Wearing a belt of snake skin that held the now bloody blade, only a patch of some kind of material covered her pelvic area. Her breasts stood out in the moonlight, the tops glowing softly in the patchy light. Long, dark hair fell across her shoulders and down her back and framed her dark eyes and young face. She was all of five feet tall, though her presence seemed much larger; the confidence and power she projected unseen in the women of the civilized world.

If I was surprised at her persona, she was just as surprised at mine. We stood in silence for a few moments until finally I cleared my throat and said, "Thank you. You saved my life."

She nodded then turned her gaze to the cart. The tilt of her head and the look on her face asking the question her voice did not; who was I and where did I come from?

"Trace," I said pointing at myself as if she did not understand English. "My name is Trace."

She laughed. "Cara," she said pointing at herself. "My name is Cara, and yes, I speak English."

Wondering if she could see me blush in the dark, I laughed a little too. "Forgive me, I meant no offense, I just did not expect to see anyone way out here."

"I have heard that people come here," she said, "but never in the dark. You are the first one I have seen with my own eyes."

"You mean tourists, people on vacation, from the hotel."

"I do not know; we are not allowed to be seen by them. Many of us sleep in the day and hunt by night."

"We?" I asked. "How many are you?"

"My people," she answered. "I do not know how many. We live here. We have always lived here."

"You live in the Everglades? In the wild?" I asked.

She nodded. "Glades," she said. "We call our home the Glades."

"The Glades," I repeated. Like the people on the island, it had not occurred to me that anyone actually lived outside of the structured communities that GOVNR had designed and managed for so long. There were not many places in the world outside of his reach. Apparently this was one of them.

40

"Can you help me?" I asked the girl.

"No, she cannot help you," a man's voice said from the trees as he approached the road where we were standing. "She has been told never to talk to strangers," the man said, glaring at Cara angrily.

"She saved my life," I explained. "If she hadn't killed that snake it would have probably killed me."

The man's look softened somewhat and I noticed that he too wore very little clothing. Not much older than me, he was covered in a loincloth made of some kind of skin; alligator I guessed; he also wore a vest, open and exposing a muscular chest and arms made from the same material. His dark hair was almost as long as the girl's and from the resemblance of his features I surmised that this must be her father.

"We are not without compassion, even to outsiders. But now we cannot allow you to leave."

From the tone of his voice and the look on his face I knew he was serious. Right hand hovering near the knife sheathed on his side, he stood poised to strike, not unlike a snake and just as deadly.

"What do you mean I can't leave? Why not?"

"Follow my daughter, I will follow you. I will explain when we get there."

"Get where?" I asked as Cara turned toward the trees and started walking away.

"Our home," he said.

A prod in the back got me moving and even though I wanted to protest, I knew had no options. The cart was dead, its power depleted from so many shuttles and my long sojourn. I would have to wait until daylight for it to power up again, I knew nothing about the Everglades and what dangers besides snakes might reside here and I had no wish to spend the night in a ditch or meet with the point of the knife that rode on the hip of the man behind me.

Moving through the brush, I hurried along the invisible path to keep Cara in sight, pushing almost unseen branches and large tropical leaves away from my face as we walked through the darkness and occasional patches of moonlight until finally coming to a stop in a small clearing.

Cara turned and pointed with her left hand. I looked but did not see what she was indicating. I looked again and still did not see anything. Finally, she took my hand and led me across the clearing until my eyes finally adjusted to the dim light of my surroundings, revealing a low, almost windowless, cinder block building. Painted in colors to match the trees behind it, I knew that even in the daylight it would be hard to see.

Her father pushing me towards it, I stumbled to the heavy wooden door, tripping on the transom and almost falling through it. Once inside I stood stock still as Cara lit a candle. No one had talked since we left the road and I waited for Cara or her father to speak first.

More candles were lit revealing the sparse furnishings of the abode and Cara's father spoke up.

"Sit," he said, indicating a small table and the three chairs tucked underneath. I pulled out the nearest one and did just that while Cara lit a fire under a large pot on top of an old stove. Soon the smell of burning wood filled the small building and now the cold gray blocks seemed warm and almost inviting.

"This is your home?" I asked, stating the obvious.

"Yes, this is our home," he answered. "My name is Hew and we are Gladesmen."

"I did not know anyone lived outside of the Communities; what are Gladesmen?" I asked.

"I have told you my name," he said. "It is customary to tell me yours."

"Of course," I said awkwardly. This was such a strange situation; first I was a captive and now I seemed to be a guest. "My name is Trace, you will have to forgive my manners but it has been a strange introduction to you and your daughter."

"Agreed," he nodded, "but you have to understand you are the first outsider we have met in decades."

I was astounded. "You mean you have seen no one from outside of the Glades for that long?"

The smoke from the fire began to burn my eyes a bit. The square block building had but one small window which swung open from the top and it did little to evacuate the wood smoke from the room, though as the fire grew hotter the smoke finally dissipated and even the heat seemed to be drawn up the slim black chimney rising up from behind the cook top.

Now the scent of burning wood was replaced by the aroma of some kind of delicious smelling food. I turned in my chair to look behind me and saw Cara at work stirring the contents of the large black pot.

"A stew," she said smiling, "and I have tea brewing too."

I smiled at the thought and my stomach rumbled in anticipation.

Cara's father cleared his throat; "As I was saying, no outsider has been in the Glades since the end of The Great War. Our grandfathers—those that survived—returned to the Glades where they were raised; many had owned businesses here before GOVNR abolished money and confiscated our land and buildings in the name of world salvation. Some of the families were Gladesmen whose roots here went back three hundred years. These were people who had learned to survive in the Glades, growing food, hunting and fishing, later running boating and tourism tours, alligator hunting and the like. Fiercely independent, The Great War was not their first battle with conservation. Before them were the native people, Indians as they were called by the white men seeking to drain this land and claim what riches they could from it. They had finally been driven out and forcefully marched for a thousand miles in the name of Manifest Destiny. Some returned later as did their descendants, mating with the Gladesmen who had not been evicted."

I saw both pride and bitterness on his face and his eyes travelled somewhere distant for a moment as Cara set handmade bowls of steaming stew in front of us. Real spoons made of metal followed and Hew saw my surprise.

"We may be isolated here, but we are not without some remnants of civilization. Our grandfathers and fathers were not ignorant savages, nor were the Indians. Much of their world and technology was hidden from GOVNR and his Techno droids. Boats, motors, books, tools and guns were moved to bunkers like this one before the takeover. Many families were killed trying to defend their homes and businesses but not a one gave in to their interrogations; the secret of those that had stayed hidden and their whereabouts died with them. Since then we have avoided all contact with the outside world. I will have nothing to do with GOVNR and that is why you cannot leave here. Now eat, your food is getting cold."

41

Though the stew was wonderful and the tea was good, the questions on the tip of my tongue kept me from enjoying the taste as I would have in other circumstances. Why would they feed me if they meant to forcefully keep me here? I wanted to trust their hospitality but like the food the circumstances it was offered in detracted from the experience. I began to look around the small household as I ate making mental notes of any possible weapons I could use.

Other questions fought for a place in line as Cara took the now empty bowl from me and I sipped on the cup of tea. Where was Cara's mother? Should I even ask? How many more families lived in these wilds? Where were the others? How could I escape or should I? Here were other people, not Biologics but real humans. Here we could live like men and women and not like Technos under GOVNR's thumb. Why not stay willingly?

I looked at Cara as she worked clearing the table, her lovely rounded breasts not hidden, her hair wild and free; so much the opposite of the women in GOVNR's tightly controlled communities. What man would not want to stay here?

But then I thought of GOVNR's plan to be rid of the human population. I thought of my friends. I thought too of all the unsuspecting families about to face extinction and knew I could not just turn my back on them. I was no hero but they deserved to be warned at least; they deserved to be given a chance.

"Hew, why did you bring me here?" I asked.

"As I said, we are not without compassion; unlike the world where you come from," he answered somewhat bitingly, his unshaven face pulling back into an angry scowl.

"You think people in the outside world are without compassion?" I replied defensively.

"They are without many human qualities," he began, "they have been mind washed by their GOVNR for so long; they live like pets in a cage."

"And how would you know?" I asked, anger rising in my voice. "You have isolated yourselves for so long wanting nothing to do with the world outside; what have you done to help your fellow man?"

"I have survived," he answered, hammering his fist on the table. "WE have survived—the last real humans on Earth."

Startled by her father's outburst, Cara walked up behind his chair, placing her hands on his shoulders she said, "Hush father, it is not this man's fault that he was born in that world."

Soothed by his daughter's words, Hew let his shoulders relax under her fingers, his scowling face as well. Nodding his head, he spoke in a calmer voice, "Forgive me, friend, my manners are severely rusted; we have not had a guest in a very long time."

It seemed Cara's hands had even eased the tension in the small block house and I too relaxed in my chair.

"And it would seem mine are rusty as well," I said, smiling. "This food is delicious and the tea is wonderful; thank you, Cara."

"Back to your question, Trace," Hew resumed the conversation. "We have a problem you may be able to help with."

I sat ready to listen, shielding my thoughts as I did so; a minute ago Hew was chiding me for being an outsider and now he was asking for my help?

"What help can I be to you?" I asked. "You have survived so long without me, as you have said."

A knowing smile crossed his face as he acknowledged my jab with a nod. "I guess I deserved that one. But all axes down; we do have need of each other."

I nodded at that. If I was ever to leave the Glades I would need their help. "So tell me," I said.

"We are dying," he said.

"Dying?" I was shocked. Never had I met two people who looked so alive.

"Not Cara and I; our people. You see, Cara was the last person born here. No child has been born since her generation. No union has been fruitful for twenty years. Whether GOVNR has found out about our existence or if something has happened to the water or food we do not know."

"Have you no doctors here? No scientists?" I asked.

Shaking his head, he replied, "No. You see, after the war, the survivors that returned were people that had made a living on the Everglades; former businessmen, boat pilots, tour guides, hunters and Gladesmen who had lived here for generations. Some knew basic remedies and cures, true, but though they were not ignorant we had no scientists or doctors among us. When GOVNR had won the war he, it, had also confiscated our land and buildings. We were forced to hide ourselves and what we could in the glades and what few outposts and buildings remained from ancient times that

GOVNR deemed not worth the resources to claim. But we knew of the satellites and the extensive network it commanded. Fearing for our lives, we have stayed hidden all these years, quietly going about our lives and shunning the outside world."

I began to see where this story was going and looked up at Cara, still standing behind her father as he reached to his shoulder to clasp her hand. Her eyes avoided mine and I could see her embarrassment.

"I fear our community is too small," Hew continued. "We have been alone too long. We need new blood."

"And you are asking for mine?"

Instead of answering, Hew pushed his chair back and stood. Glancing at Cara with a slight nod, he straightened his vest, touched his side to check the presence of his knife then turned to the door. "We need more meat," he said. "I'll be back later."

The heavy wood door pulled tight from the outside as he exited the small camouflaged block house leaving me alone with his daughter.

42

I don't know how much time passed before Hew returned that night. I do remember Cara's beasts against my chest and getting lost in the long dark hair that fell softly around my face as all the cares of the world rolled off my shoulders and fell to the floor with my unconvincing protests.

I woke in her bed in the morning to the aroma of some kind of meat emanating from a pan on the old stove, filling the little house with a smell that seemed to link directly to my rumbling stomach.

As I sat up on the bed, I noticed I was naked and pulled the blanket across my lap, embarrassed. Though Cara was at the stove humming some kind of tune, I was relieved to see that her father lay still asleep.

I looked around for the clothes I had worn, donning them one by one as I scooped them from the floor. I had not known these people long enough to be sitting here like this or to have done what I had last night. The curious mix of contrasting emotions was new to me; so unlike the analog of feelings that had matched the events around me from the time I had left the Med Center.

Cara had turned from her cooking now and then, watching unabashed as I had dressed, apparently amused as my face reddened with each glance.

"Good morning," she said smiling. "You can pee outside before breakfast."

Again her frankness startled me, but I stood and followed the sudden urge to get outside. I was not sure I wanted to come back in.

The warm humid air enveloped me as I searched for a spot along the edge of the small clearing. Even though the sun had not been up for long, its heat seemed magnified by the water that permeated the Glades. As I stood looking into the trees, I struggled to organize my waking thoughts. I had been occupied with one single purpose for so long; to find out who I was and why I seemed to be a threat to GOVNR and his world. In all that time I had learned only that I was a Sleeper; someone from the past who had been cryogenically frozen to be brought back at some point in the future, presumably when I could be cured or repaired of whatever had ailed me. If that were the case, then was my "fall" something that happened in this time or could it have been the catalyst for the cryogenics? I still did not know and staying here would not help me find the answer: Beyond that was the larger concern; GOVNR's plan to systematically erase all humans from existence. That trumped my petty little problems and gave me all the more reason to leave this place; if I could.

Without a sound to announce his presence, Hew suddenly appeared beside me. "Not thinking of leaving us I hope," he said.

Startled out of my thoughts, I jumped at the sound of his voice as if I had been thinking of just that.

"As a matter of fact I was," I said. "We are all in danger; GOVNR plans to wipe out humanity."

"And your plan is to run?" Hew asked as I finished and faced him.

"No, I am through running. I have seen him at work; the last place I stayed was burned to the ground, the people killed as they ran. If I stayed here it could happen to you and your people too."

"Why should I believe you?" he asked. "How do I know you are telling the truth?"

"What has happened to your people? Did you not tell me they were dying out? You said you had managed to stay hidden all these years but how do you know you have? Do you not think GOVNR would know there would be survivors from the war and that they would return here and that they would be rebels?"

"That is why you cannot leave," he answered, spitting on the ground before continuing, "That is why we could not let the others leave either."

"There were others?"

"Yes," he admitted, "when we first realized something was wrong and no women were getting pregnant, we assumed that our gene pool was too small. Occasionally tourists would wander in from the Hotel and some of them would, 'disappear.' We offered each one the same choice; if there was a baby born stay here and make more; raise the children and learn our ways."

"And if no child was born or they chose to leave?"

"Well, then we could not chance them giving us away. They would quietly disappear. Accidents, the Hotel staff would say, got too close to the Gators…"

He had finally voiced what I had been thinking since our first encounter and the trek to his house at knife point. At this moment in time I wondered if he were any better than GOVNR himself.

"You're killers," I said. "No different than the one you are hiding from; anything to survive, right?" I spat on the ground in front of him. "And just where are your people? I have not seen anyone but you and Cara since I arrived."

"What are you saying?" he asked. "That there are no people here?"

"Why haven't I met any of them? I don't see any other houses, camouflaged or not."

"They stay hidden," he said, looking away as he answered. "Away from each other we have a better chance of not being discovered."

Something about his manner raised the hair on the back of my neck. Things were not what they seemed to be. Now the training I had received from the Underground began to awaken in me, and the time I had spent on the bogus missions they had sent me on began to serve me as I realized what had set off the alarm.

Though sweat trickled from my brow as the sun rose higher in the morning sky, Hew looked like he hardly noticed the heat and humidity at all. And while he and Cara may have been used to living in these conditions, neither one gave off any body odor. Not even bad breath.

Of course! I did not know why I hadn't seen it sooner. They were Technos! If I had been nervous before, I was doubly nervous now. What was their game? Why the charade? Should I confront them or play along?

"Discovered by whom?" I asked hoping that if I let him talk enough he would tell me more than he wanted to.

"Why, GOVNR of course," he replied. "He took away our world, our way of life—that is to say he..." now it was Hew that sounded shaken.

"He what?" I prodded.

Cara's voice interrupted the awkward conversation. "Hey, breakfast is on," she called from the door. "Are you guys hungry or not?"

Not, I was thinking of Hew. But I knew I was. Whatever Cara had prepared I was ready to eat; I would need the food if I hoped to leave here.

Once inside at the table, I sat down to a plate of meat. And though there was a plate in front of Hew as well, I saw none for Cara.

"Oh, I ate while you two were talking," she said, bending down to let her bare breasts brush into me as she set a cup of tea next to my plate. Though they may not have been Biologics, they were very convincing. But the thought that they were Technos helped minimize the distraction as I picked up the fork and dug in to the still steaming fare.

"Hew," I said between bites, "when can I meet the other people?" I asked as if I intended on staying.

That brought an uncomfortable silence into the small block house and a puzzled look to the face of my host and captor. Cara too had stopped as she cleared the food from the old stove.

"You can't," he finally answered.

"Why not?"

"They are all gone now," he said.

"Gone? Where?"

"Not where; when," he said pushing his plate away before continuing. "It was our job; we were born—created—to infiltrate the human survivors of the war that would return here. One by one we gained their trust and then eliminated them. GOVNR had declared victory and an end to the war. He promised not to terminate any of the humans that had survived. That is why he renamed them Biologics. Then he ordered that pockets of rebels be hunted down and neutralized. That is what our job was."

"Neutralized; such a nice word for murder," I said acidly as I pushed my plate away. I was suddenly not hungry anymore. "But that was long ago, have you and Cara been here that long?"

"Yes. When the task was completed, GOVNR cut all ties with us. He left us here to rot."

"But you could have gone anywhere you wanted to," I said. "Why did you stay?"

Shaking his head, Hew looked me in the eyes. "You are wrong; GOVNR would not let us leave. He threatened to terminate us if we did."

"So you kept up the charade," I said. "Kept on killing. Why? Were you trying to win back his favor?"

"We just did not know what else to do. We were not gifted with the intricate processors of the later generations. We were designed for one purpose. We did not want to keep killing but when the vacationers would wander too far and discover us, we had to do something. We did not want to become a problem that GOVNR might solve."

"I am not moved," I said quietly. "You chose to kill to save your own tin hides; you are cowards! I am disgusted and I am leaving!"

My chair flew back as I stood up angrily and if looks could kill the one I gave Cara should have been lethal. But Hew stood up as well and blocked my path to the door.

"You are not going anywhere," he said as he drew the knife from his belt.

I might have been physically fit but I was no match for a Techno; especially one that was designed to kill. If I couldn't match Hew with brawn maybe I could match him in brain.

I turned away from Hew and his knife and looked at Cara. "If you were going to kill me anyway, why not just let that snake do it back on the road? Why put me through all of this?"

"Like my father told you," she said still playing her role of daughter, "we are not without compassion and even we get lonely."

I had to admit she sounded human—like a human psychopath maybe—with her feigned emotion and the drooping of her shoulders designed to lead my eyes to her quite human looking breasts. But I could not let myself be deceived and let emotion get in the way of the facts. And the fact was they were both killers.

Self-preservation is a powerful force; even for Technos. And it rose up in me just then too.

"Wait Hew," I said, raising my hands for emphasis. "I'm sure you are good at what you do…"

He nodded. "Very good," he said with almost human-like pride.

"Then why just do it here where it is so easy; why not give me a sporting chance? Let me get out the door, give me a head start—say thirty minutes—then come after me. You know the glades so well, I'm sure you could track me down easily enough."

"It would a quicker death for you here," he said. "We both know the glades well as you have said and you would just be prolonging the inevitable."

"But at least I could die with some pride," I said as I looked unflinchingly into his eyes.

"A Biologic with pride, I have not encountered one for a long time. It is your pride that created this world; your pride that brought us about. We are a product of your arrogance and it has cost you your world," he said.

"And your arrogance is showing now," I said. "How about it? What have you got to lose but a little time?"

"Time we have," he agreed. "Very well, human, the challenge is accepted but to have a real chance we will grant you sixty minutes."

I was not going to argue. I would need all the time I could get.

43

There was one thing I had that Hew and Cara had not counted on; something that had been lost in the hundreds of years of what was once called artificial intelligence—resourcefulness. With each advance throughout the history of mankind came a price; one that was extracted each time a tool was created that made life a little easier. And as those advances piled up, so did the cost until finally the compound interest surpassed our ability to pay and we found ourselves mired in debt, our resourcefulness all spent.

And when that was gone we dug a little deeper into our pockets and began paying with our freedom and finally our humanity. Now our new masters would not be content with anything less than our lives and ultimately our extinction. We had bargained away the world we had been given.

GOVNR and the Technos were well versed in our history but knowing a thing and understanding it are two different things. Hew and Cara had never had to be resourceful. Their job was to infiltrate, deceive and eliminate the rebels and survivors in the Everglades. Specifically designed for the task, they could mimic people, feign emotion, hunt and kill but they lacked the very thing that humans were born with, the very thing that had led to their creation; resourcefulness.

Though I still had no memory of my life before the Med Center, the knowledge that I had been a "Sleeper" had begun a process that the whole previous year's events had failed to awaken. I had begun to get in touch with myself; my real self, not the artificial man created by GOVNR's world. And the events of the last few days had accelerated the process.

True, I knew nothing about the glades and I was no match for Hew or Cara but I sensed that my will to survive was greater than their will to end my life. And I was counting on that.

"Your time starts now, human," Hew said as he moved away from the door.

While the two killer Technos had an accurate internal clock, I knew I could only guess at the passing of time and I wasn't going to waste any on goodbyes. Without a backward glance, I lunged for the door, determined to put as much distance between the camouflaged block house and myself as possible.

Once outside I headed for the trees directly ahead of me oblivious to the wild bushes and exotic fauna I was crashing through. There was no time to step carefully or attempt to cover my tracks and as I ran into the unknown a plan began to form in my mind.

From Hew's comments on the shortcomings of mankind, I was pretty sure they would split up and follow two different paths; one was the trail that had led us to the house from the surrey topped hotel cart on the road where Cara had found me and the other was the spot where Hew had stood with me before breakfast. It was logical to assume I would head for one of the only two places I knew outside the little house. And I was pretty sure they would assume that I was like the other people who had been unlucky enough to wander too far from the hotel tour; Biologics with no imagination.

Intuition was telling me that my best chance of escape was what they did not know about me. I needed to be unpredictable. I knew it was likely that they would eventually find me and I also knew that I was not big enough or strong enough to bring them down. Not without help anyway. And to get that help I needed all the time my one-hour head start would buy me.

44

After my initial bolt from the cinder block house into the trees, I willed myself to slow down; partly because of the bushes that hit back as I crashed through them and partly because my brain was trying to crash through the adrenaline rush and my pounding heart.

Though I felt more alive running for my life than I had in the whole previous year; more alive than the recent sex had made me feel, the thinking part of my mind was telling me to go slow, walk soft and think.

Distance did not equal safety. Technos were capable of great speed even in rough terrain, especially familiar territory. And their senses were sharper than ours; their ears capable of tuning in to any unusual noise and sending it to their processors at lightning speed and zeroing in their target. No, distance alone would not save me.

I knew that at some point both Hew and Cara would catch up with me. What I needed was time to prepare. If my hastily formed plan worked I could take out one of them but I would not be able to repeat it. I would have to improvise when the time came.

Slowed to a stealthy walk through the brush, I stepped carefully in the soft, marshy ground, lengthening my stride to cross rivulets of water as I pushed west toward the larger body I somehow knew was out there. Looking back over my shoulder, I could see that my tracks would not be a problem. The high-water level rose to fill them as my feet plodded through the soft ground with a shucking sound and my legs tired with the effort of pulling them free from the greedy mud that did not want to let go.

Higher patches of drier ground rose up here and there as I travelled, my mind busy both navigating the wilderness and visualizing the coming confrontation. If I read Hew correctly, he would follow my path into the trees telling Cara to backtrack to the road where she had found me. I was counting on that to buy me enough time to take him out before I had to deal with her.

I grew more certain of my plan with every step and was now navigating on auto-pilot. Whether instinct or desperation, I no longer cared as my fear transformed to rage. Who were these machines to have the power of life and death over their creators? And how had it come about that humans had ceded them the deed to the world? It was our arrogance that had put me in this situation and now I was counting on that same arrogance in the Technos to save me.

Suddenly aware of the heat and the late morning sun, I found myself on the edge of a large body of water looking out on several large crocodilians moving towards the shore, their bumpy hides just visible on the surface of the water. To my right were several more of the large reptiles apparently sunning themselves on the silty shore.

Standing quietly so as to not disturb them, I felt the sweat running down from my armpits and realized that the animals had picked up the scent. I looked at the mangrove trees for one I could climb. Fortunately, I was not far from a grove of them and the roots along the water's edge would enable me to climb into the sheltering branches.

All I had to do now was get past the bumpy bathing beauties.

It would be close to another hour before Hew found me and he was helped by the large birds that flew out from the mangrove tree as I settled into a crook partially hidden by the large sheltering leaves.

My dash across the field of sunning crocs unaccustomed to the smell of a human had left them hungry and restless and milling around on the narrow, silt covered strip of sand at the water's edge, threatening to give away my hiding place; which was exactly what I had been hoping for as I watched Hew come into view.

Long ago these crocodiles had been known to be a little skittish but over the last two hundred years things had changed. With no humans to hunt them or disturb their habitat, their numbers and boldness had increased and it was only my speed that had prevented them from making me into a lunchtime snack.

Hew stood, apparently unafraid of the predators as he scanned the water, I assume calculating my slim chances of survival. If I had tried to swim to the other side, I would surely be dead and eaten or at least bitten and bleeding by now. Judging by the animal's restlessness, I had been here not long before.

Scanning the opposite shore and then the water again, and finding no sign that I had attempted to cross, the Techno turned his attention to the trees.

Since I was no match for him physically, I knew that my only chance was to get him off balance and I had positioned myself to do just that; choosing a branch that hung dangerously near the water's edge. I stayed perfectly still as Hew looked up from the crocs to the trees using his enhanced vision to see through my camouflage.

Wasting no energy on words, he began his trek through the gathering of reptiles, unconcerned about being bitten as they regarded him with curiosity. He had no scent. But I did and my scent had aroused their appetite.

My heart began to race and I was certain that Hew's ears could hear it as it pounded faster and harder against my chest. All the better, I thought as he moved into position in front of the tree deciding on a course of action. Should he throw his knife or climb up and pull me down?

I was not sure which predator was more deadly as I readied myself. I would get one chance at this and no more.

45

But one chance was all I needed. As my enemy stood there scanning my location, the sweat from my body and the scent of human flesh brought several of the crocodiles to Hew and when he looked down to see himself surrounded, I grabbed the branch in front of me and swung with all the pent up anger and adrenaline I had, pointing both of my feet at his head and shoulders. They connected with enough force to knock him off his feet, sending him backward over the back of one of the hungry crocs.

Now on top of the killer Techno, I knew I had little time before the surprised animals descended on me. I reached for the knife in Hew's belt and once in my hand, I aimed the point at his right eye and drove it in. Before he could react ,I stabbed him in the other eye and rendered him blind.

But I knew I had only seconds to scramble off of him and back up the tree to safety as he strove to fend off the now angry crocs and the snapping jaws that pierced his artificial skin and tore at his limbs.

The weight of the huge animals and his blindness was too much for even his machine-like strength and I watched as three of the predators fought for their prize, tearing large pieces of nano skin from his frame and dragging him helplessly to the water.

I felt no pity for him as the largest of the crocs went into a death roll with the once killer machine, pulling him under the surface until both the victim and the water were finally still.

One down and one to go.

Whether Hew and Cara were connected somehow or he had sent some kind of signal to her, I did not know but it was not long before I heard a crashing in the brush and saw Cara run onto the beach.

The milling of the crocs, still agitated by the recent struggle, caused her to slow her pace and tread carefully along the edge as her mind processed the probable scenarios. Looking down at the sand, she read the footsteps and tell-tale signs left by both Hew and the crocodiles and as I watched from the safety of the sheltering branches around me I willed my heart to slow, becoming one with the supporting branch I was leaning on.

I doubted that the same plan would work twice and I did not want to press my luck with the hungry beasts below. But I still had Hew's knife and so far Cara had not sensed me.

The few crocs that watched me scramble up the tree were still milling about below me and though their short necks did not allow them to give away my position, I feared Cara would look up and spot me. Fortunately, the lumbering giants turned their attention to her and began moving in her direction.

I had no doubt that one on one, Cara could defeat them, but with three of them all moving toward her at one time she started to back up slowly, keeping her eyes on the closest one as she backed into the brush, and out of harm's way.

I knew the crocs could move fast for short distances but I doubted they could catch her if she turned and ran with the great speed I knew she possessed. And that is exactly what she did.

From my vantage point I could see beyond the low tangle of tree bush and as the lead crocodile gave up the chase at the edge of the brush, I watched Cara speed away back towards the small camouflaged house as I pondered my next move.

I wondered how long I would have to stay in the tree before the crocs wearied of the wait. Still hungry and aroused by my human scent, I hoped that other food would appear in the water and draw them away.

46

The time of day must have had something to do with their eventual withdraw and as the sun continued on its westward journey, I looked out from the sheltering leaves of the tree to see an open stretch of sand and quiet waters and began to move my stiffened body out of the clutch of the two large branches that had held me for so long.

The long wait had been put to good use and I now had a plan; a dangerous one maybe but a plan nonetheless. And as I moved across the sand I knew that Cara too would have a plan.

Both of us knew the same thing; that I would have to get past her one way or the other if I ever hoped to leave the Glades—if I was indeed still alive. With no way to confirm the kill, Cara would wait at least a day before declaring me dead and when Hew did not return in that same amount of time she knew he would be dead too. My lack of experience would tell her that hunger would draw me back to the little house where she would undoubtedly be waiting and that in desperation I would try to bargain for my life.

And she would be right. Hunger would drive me back but not the way she anticipated.

The one thing I had going for me was the element of uncertainty and the arrogance of the two killer Technos that had successfully erased all human occupants of the Everglades. In the short time they had known me they would have made assumptions based on their interaction with the people of their own time; of which I was not.

Somewhat confident in my rationale, I moved quietly through the brush and into the clearing beyond, setting a path for the little house hidden in the trees beyond.

Technically speaking, I was guilty of murder even though I had killed Hew in self-defense. In this world they called it termination; something that supposedly had not been done in years. But I knew better. Hew and Cara had been designed and planted for the sole purpose of killing any Biologics who crossed their path.

Designed, now there was a word that made it hard to think of them as sentient beings. And GOVNR, though a conscious entity and technically alive, seemed more like an abomination than a living being. But why should I be surprised at any of this? Artificial intelligence had been born by humans, raised by the military and the establishment, fed the whole of human history and set loose to run the world. How could we expect it to be any less devious than its creators?

And instead of being revered, we had shunned them, excluded them and refused to treat them as equals. Regarded with prejudice, treated like slaves, abused like machines, we continued to treat them as unthinking robots incapable of feelings or a sense of purpose. And after decades of caretaking, it came as no surprise to me that they would tire of our presence altogether.

It was no surprise how quickly it happened.

The rage I had felt when I had attacked Hew was not just for him and Cara or even GOVNR and his attack on St. Croix Island. Nor was it just the deceptions of the Underground and my frustrated efforts to find my former self. It was the whole stinking world, past and present and the power-hungry greed and vanity that had fueled mankind's inhumanity since the dawn of time.

And the same rage was alive in me now as I approached the camouflaged house in the clearing at the edge of the wood; a primal rage driven by the need to kill or be killed that both heated my mind and chilled my heart at the same time.

A Techno's greatest asset, its eyes, were also their greatest weakness. Though many of them were created with different "gifts" according to the jobs they were to fill, most relied on their eyes more than the other sensors they were equipped with. And like Hew, it was Cara's eyes that I needed to disable first. But this time I had no four legged allies to help me.

Unlike humans, who all shared the same skeleton, the Technos differed in their construction and materials. The first and oldest ones were made with Titanium frames but as their numbers increased and the metal grew scarce, the use of super strong carbon nano tubes increased. Over the years their power sources had changed as well; from hydrogen to solar and hydrogen/solar hybrids all according to GOVNR's allocations of resources and the tasks the Technos were charged with. Some, like the Security team that acted as police were specially equipped for possible confrontations and I had no doubt that Hew and Cara had been equipped with multiple defense mechanisms beyond their great strength and speed.

All those thoughts served to channel my rage into a risky and deadly plan and as the house came into view I looked around for the weapons I would need.

47

As with Hew, my greatest weapons would be surprise and rage. Of course, a few large rocks and a fairly strong sack to load them in would help. Hew's knife would deal the final blow if Cara was indeed inside the small concrete block house.

Not finding any suitable leaves to make some kind of a sack with, I settled on collecting a rock large enough as I moved carefully through the tangled bramble at the edge of the clearing. Finding one big enough to do the job but not too big to carry, I circled the clearing in front of the block house to the trees behind it.

The next challenge would be to climb the tree behind the house and get on the flat roof. I knew I would not have much time as Cara's hearing was probably very good and she would hear me moving overhead. Climbing with a twenty-pound rock in one hand was not easy but my heart was pumping against my ribcage and I knew adrenaline would give me the strength I needed.

Somehow I made it to the pitched and painted blocks of the little house and crouched in wait, catching my breath and willing my heart to slow. There was no parapet around the roof and crouching or even lying prone would not hide me. If Cara was not inside, she would easily spot me from any angle of approach. No matter; all I needed was for her to get close enough. And if she was inside, all I had to do was wait for the door to open.

Again, I would only have one chance at this. But my position on the roof would give me an advantage and so I waited, envisioning my attack and running every scenario in my head until the daylight began to wane.

I had to act soon; darkness would not be my friend. There was no sign of Cara as I scanned the area and I reasoned she must be inside, probably calculating the odds of Hew or myself being alive. Perhaps the hungry crocs had not eaten us after all or maybe just one of us.

Cursing the fading light, I stood up into the dusk as it slowly settled over the house and stomped my right foot. Once would not be enough as it could have been an animal so I reasoned that three hard stomps would draw her outside.

I waited and listened.

And then I heard the large wooden door open.

I watched as the top edge of the door swung into view, rock poised above my head ready to strike. But no one stepped out. She must have opened the door and waited inside, scanning the clearing.

The weight of the rock began to weigh on my arms and my shoulder muscles were beginning to burn. I realized I was holding my breath and knew that I could not do it much longer.

Finally, Cara stepped through and my aching arms pitched the small boulder down on her head, knocking it into a weird sideways angle and causing her to struggle to keep her balance; a struggle she lost as she fell on her left side, dazed and confused.

I wasted no time and jumped the short distance from the roof, rolling to take the shock off my ankles and landing a few feet in front of the killer Techno. I was up and on her in a flash, Hew's knife aimed for her left eye.

Though she was indeed damaged, she was not out yet and the killer mechanisms she had been made with brought her arm up at lightning speed to block my thrust. Instinctively, I tightened my grip on the knife, managing to hold onto it as I rolled away from her block. A quick sidestep brought me over her body as she fought to right herself and regain her feet. Her head still hung at an unnatural angle though and I knew her balance had been damaged.

Now attacking from the other side, I aimed the point of the knife at her other eye and again she blocked my thrust, this time knocking me several feet away with her inhuman strength.

The blow had sent me reeling and I knew I would fall.

The knife had flown from my hand and vanished into the spreading darkness of the early evening and the shadows cast by the house and I came down painfully to the ground, my ribs contacting something round and hard.

The rock I had struck her with now lay under me. I hoped my ribs would just be bruised instead of broken and I gathered the rock into my hands, pushing myself up on it as I stood with my back to the injured Cara.

One more chance, I thought to myself as I turned and ran the few steps toward her, rage and adrenaline pumping through my veins as I raised the rock and brought it down on her head again and again until her mangled skull was finally still.

And as the night descended, her body disappeared into the darkness and I sat exhausted for a long time before finally rising and stumbling into the house that was now mine alone.

48

I don't know how long I slept but when I awoke and opened the door, daylight flooded in. It was sometime late in the morning, not yet noon I surmised as the sun was just peeking over the trees in front of the clearing.

I looked down to see Cara's body. No longer hidden in the dark, her bare breasts seemed almost ghoulish under the mangled head above them. I felt no remorse. Sentient or not, she was not human. A robot mimicking life. I turned from the scene and went back inside.

There wasn't much in the way of food on the shelf above the old stove. Apparently, the only time they cooked was when some poor, hapless Biologic fell into their evil clutches. Some zip-foiled beans from one of the Techno farms were all I managed to find. The same ones they used for their snake meat stew. But I took them anyway along with whatever of Hew's few clothes I could find and anything that I could use as a weapon; an antique hammer, an equally antique hatchet and the trusty knife that Hew used to carry.

Sure, I could probably have stayed there and learned to survive but it would have been a lonely life and eventually would drive me to kidnap tourists. Then there was the matter of GOVNR's plan to finally eliminate us bothersome Biologics…

I could not stand by and do nothing even if my efforts led to futility. Nor would staying here answer the questions about my past and who I had been before I found myself in the Restaurant serving the slop they called food.

I had grown so accustomed to the recurring pain in my head that it no longer registered much. The events of the last few days had all but eliminated its power over me. I found that if I stayed in motion keeping my thoughts away from my past that the pain lessened and I began to believe it was caused by a memory block; probably one inserted into every "Sleeper" that woke to a new future as well as being a tracking link to GOVNR.

And that reminded me that GOVNR had probably closed the file on me after destroying the island I had escaped from, which led me to the basis of a new plan beginning to take shape in my brain.

I would head back to the hotel and rejoin civilization.

Tools and beans stuffed into a snakeskin pouch and wearing one of Hew's snakeskin vests, I left the concrete bunker and started down the path to the overturned tourist cart back on the road.

49

The cart proved too heavy for me to right it. The two wheels in the ditch were at such an angle that the full weight of the vehicle fell on my overworked arms and shoulders and after a few attempts I gave up in exhaustion and slumped to my buttocks on the roadside of the useless machine.

A Techno could probably have done it with minimal effort but a poor Biologic like me would need a very long lever. I entertained that thought for a few moments while scanning the ditch and the trees for something I could use but even if I had found something strong enough to do the job it would have to be very long to make up for my lack of weight.

So I decided to walk. It was forty miles but I had wanted to conserve my energy. I knew when I did return to the hotel that the only occupants would be Technos and no doubt would have me on the security scans even if they did not know my identity. I could only hope GOVNR would not be alerted if they did.

My plan had been to drive the cart back, report an accident and be assigned passage on any remaining shuttle to anywhere. Or sneak aboard one. Not a great plan but it was all I had at the time.

The long walk did have some advantages; it would give me time to run scenarios and work up a better plan and it would put me closer to nightfall and the cover of darkness. I started to like both options and pushed off the side of the cart as I stood up and retrieved the waiting bag of supplies before setting off down the road at an unhurried pace.

But all the morning workouts on the treadmill soon increased the speed of my feet until I was covering the distance faster than I realized and I found myself jogging along the dark ribbon listening to the sound of my strides become like the ticking of seconds on an ancient clock; one that ticked with a sense of urgency as if time itself was running out.

The ticking grew louder as it flowed with my blood from my feet to my legs, my heart and my brain, infecting me with that same sense of urgency and dread; time was indeed running out. GOVNR was calling everyone home; Phase Three must have begun and whatever it was I knew in my veins that it was not a good thing for the Biologics.

Urgency and dread soon turned to fear and rage as I raced through the fading heat of the day, oblivious to the sweat on my skin and the fact that I had no plan whatsoever as I approached the hotel with an hour or more before sunset. But this was GOVNR's world and darkness would not have cloaked me anyway.

Maybe the best plan was to have no plan; GOVNR had two hundred years of creating a perfectly managed world filled with passive and predictable Biologics and Technos that he could turn on and off at will. There had been few surprises for a long, long time; and I was determined to be one of them.

I had slowed my pace to a walk as I came to the long line of empty carts past the Drop Off zone: Where once they seemed to wait expectantly, they now looked as if they had taken their last ride.

The hotel itself looked as if it had died. There were no signs of life, no tourists on the lawn or the walkways, no one on the beach, no Techno waiters pampering the denizens of the sculpted castles in the sand. Even the water was still.

Wary of any watchful eyes, I walked with trepidation, following the cart path to the front of the large white building and the ominous double doors, now closed and seeming very unfriendly.

Surprised to find them unlocked when I pulled on the oversized handles, I braced myself for a confrontation. But I was met by only a silence as quiet as the plush carpeting under my feet.

Instinctively, I knew that this place was empty and that caused a shiver to travel up my spine, adding more speed to my racing heart as I padded across the carpet from space to space, passing couches and chairs that would never grace my small, bland plastic apartment. I could almost hear the ghosts of the many conversations that once filled the giant lobby as the vacationers filed in and out, anxious to rest their backsides on the luxurious cloth or run barefoot into the gentle surf.

Walking past the check in desk, I wondered why they had all gone. I knew that they had been sent to their home cities but I did not know why; and that why filled me with dread. Of course, GOVNR was now activating Phase Three of its plan and though I was not certain of the details, I had a strong feeling that time had finally run out for us hapless Biologics.

Reaching the kitchen, I was suddenly aware of my rumbling stomach and the beans I had carried in my sack. I began to search the well-ordered lockers for something more appealing to fill it with.

I was fortunate to find most of the stores were left behind and went to work on preparing a meal that would have fed two hungry guests.

The sole diner in the large banquet room, I watched the sun go down and wondered what I should do next.

50

The hotel had its own cars and it was not difficult to find the I.D. cards to operate them. I chose a large one that was meant for driving important officials and loaded it with food and water.

As much as I wanted to stay the night and sleep in one of the giant beds, I knew I had to return to my home community and warn my friends of the danger I was sure was coming. I told the car to map my destination and it dutifully complied, driving itself out of the underground parking area and onto the main road, its powerful lamps lighting the way ahead while the soft voice of its processor told me I could sleep and it would wake me upon arrival in the morning and as the dark shapes of the landscape flew past the view ports, I slipped into dreams as disturbing as the shadowy images outside the car.

Automatically filtered, the morning light woke me before we arrived and I worked my seat back to a sitting position as the car informed me that we were now one half an hour from our destination. Rummaging through the supplies I had brought, I found a muffin and some water for a quick breakfast and informed the car to stop so I could relieve myself.

While I stood partially shielded by the vehicle, I looked around for any patrols or cameras that may be watching my arrival. I still had no plan and decided that fact was my best course of action. There were too many unknowns to have done anything else and climbing back into the waiting car, I instructed it to head right down Main Street and take me to my apartment. If it still was mine. I needed a shower and a change of clothes.

Leaving the car to wait for me, I walked into my building as if it was just another ordinary day. There were no Technos to be seen and unchallenged once again I walked to my door, determined not to let my apprehension show. The door slid open and I stepped inside, undressing as I went and stepped into the shower, grateful to have come this far.

The walls were silent as I dressed in the uniform of the day and for the first time appreciating the feel of the clean clothes I had once despised. A new pair of shoes for my feet and my outfit was complete; a clean white sheep ready to be sheared.

Having grown used to silence on my recent adventure, it took a while for me to realize that something was not right.

"Cyber," I said, "show me the news."

There was no response. No screen appeared on the wall and the once familiar voice that used to greet me remained absent.

"News," I commanded again. And again there was no response. "Will you talk to me at all?" I asked.

No. Well, if I had been declared dead then this living space would be reserved for someone else, I reasoned; someone not assigned yet. Essential services were functional like the water and lights but since each space was programmed for its occupant, other services would not be.

This thought gave me a feeling of power. If GOVNR assumed me dead, I would have more freedom of movement. But I still had no access to vital information and the tablet I carried had no connection to any network.

I would have to rely on my friends, unless…

The hotel car! It was functional and should have access to the network! I hurried to the door of the unresponsive apartment and out onto the street to the waiting car.

But the space I was sure I had left it in was empty. The car was gone.

51

Staring at the vacant parking space, I became wary and self-conscious. Though I was dressed, I suddenly felt naked. Silent alarms were going off in my head and I knew I was being watched. But that was normal here. The reality was more sinister; not just watched but manipulated.

Of course, I should have known that when the door to my apartment opened for me but the apartment itself was uncommunicative. Not only that, but the street was as empty as the parking lot I stood staring at. Not a Techno or Biologic in sight.

I began to question my senses. Had my recent experiences somehow changed my perception so that the once familiar place I knew now seemed alien and ominous or was something going on? Was I too late? Had GOVNR already carried out Phase Three of his plan?

One thing I was now certain of; I was not off the rolls of the living and as such any advantage I would have gained from being "dead" was gone and like a runaway commuter train it hit me; the very clothes I wore and the sidewalk under my feet were transmitting all of my bio-functions and emotions directly to GOVNR.

With nothing to lose, I turned and walked as calmly as I could towards the Park. Again I moved about unchallenged as I passed the apartment buildings that lined the street, each one as lifeless as the artificial grass of the boulevard.

As the sun moved closer to midday, I neared the Park, noting the lack of foot traffic that usually signaled the zombie-like exodus to the Restaurant. Where was everyone? Why was I the only person here?

No signs of calamity shown anywhere; nothing looked out of place and the trees in the Park hid no bodies and answered no questions. But the trek had brought me closer to my friend's house and I turned in that direction, worried at what I would find when I got there.

The walk through the Family section was no different than along my street. The houses that once held living souls looked haunted even under the afternoon sun, increasing the growing sense of dread that I felt and the apprehension about the fate of my friends.

The door that once greeted me remained mute and would not open. I walked around the dwelling to the patio where I had shared many hours with Art and his family. No one greeted me as I walked onto the recognition blocks; no colored lights appeared under my feet to announce my arrival and peering through the glass doors I saw nothing but the empty kitchen and garden room.

What would happen if I broke the glass? Probably nothing, I thought. Crime was an antiquated term in this world. But what would I use? I carried no weapon and the plastic chairs around the patio table would be useless against the thermal glass. I looked around for a large rock but the ones here were ornamental—unlike the one I had used to kill the psychotic Techno in the Everglades.

The trees were as phony as the rocks but I looked at the limbs for one I might be able to break off and use as a club. I chose one and reached up to break it off.

"They're gone," a familiar voice startled me out of my concentration and I froze, my arms extended up towards the branch. "They are all gone," Celes informed me.

"Gone where?" I demanded.

"Just gone," he replied as I lowered my arms. "When I returned from the island after dropping you into the water this is how I found the whole town."

"And you have no idea what happened?" I asked suspiciously. Technos could communicate by thought, much like wireless networks and only spoke when addressing Biologics.

"I have been cut from communication with GOVNR; apparently in response to the help I had given you. I have been isolated from my own kind and denied access to any network or vehicle. My team has vanished as well and probability suggests they are now under GOVNR's command."

I looked deep into his eyes as if he were a Biologic; they told me nothing. How could I trust him? I was not even sure I could trust my own senses.

"I see doubt in your eyes, Trace. Why would I deceive you?"

"You would not be the first," I answered. "You are a Techno."

"Yet here we are," he responded, "both of us standing in an empty place not knowing where the inhabitants have gone. The people we both knew and the ones I protected. Only GOVNR could have done this."

I had to agree. This must be part of Phase Three, I thought. I wondered if Celes knew anything of it.

"What is Phase Three?" I asked.

"I have no knowledge of Phase Three," he replied. "Phase three of what?"

"The plan," I said. "GOVNR has a plan and I think it is to eliminate all of the Biologics once and for all."

"How did you learn of this?"

"That is not important," I said not sure if I could trust him. For that matter how did I know I wasn't talking to GOVNR himself? And I felt a need to protect Pral and any of the people left on the island. "What I need to know," I said, looking into his artificial eyes, "is if you have any knowledge of such a plan."

Biologics had no reasons to lie in this world. There was no status to achieve, no wealth to hoard, no land to own and most of them had been spoon fed by GOVNR for generations. One could easily spot deception in people that were so unaccustomed to using it. But Technos were not Biologic. I watched his eyes closely.

"Why would GOVNR share such knowledge with me?" he asked.

Was he avoiding an answer? His eyes, programmed to show any appropriate response, told me nothing.

"Because," I replied, "you are the Security Officer here. You would have had a need to know."

"I was the Security Officer," he corrected. "I cared for all the residents, Techno and Biologic."

"Equally?" I asked.

"Yes, equally. It was my job." He stared back into my eyes. "It is you who harbor mistrust, not I."

"Then you did not take the car from my building this morning?"

"I was not aware you had one to use," he replied, a note of authority still lingering in his voice.

"If you did not take it, then who did?" I asked.

"It may have simply been recalled or set to return to base…"

"Or GOVNR could have done it." I finished the sentence.

Celes nodded. Those were the only possibilities, given the fact that there were no people anywhere to be seen.

"Which brings us to where we are now," I said. "Where is everyone and why have they gone?"

"And why would I not know about it? It seems you may be correct about your Phase Three theory. If that is the case, then GOVNR must have called all of the Technos as well."

"All except you," I pointed out, suspicion still clinging to my words. Perhaps he was here to collect me.

"What will it take to convince you that I was not sent here to pick you up?" Celes asked sounding very human now.

"Help me find the others and free them, if it is not too late," I answered.

52

Did I know I could be walking right into a trap? Of course I did. But what else was there to do? The fact was that Celes could have overpowered me anytime he wanted to. There would be no need for subterfuge. It was only my deep-rooted resentment that drove my suspicion. In a world run by machines the only thing that did not make sense was me; and the fact that I was working with one now.

"I would suggest we start our search at the Government Center," Celes said as we walked away from Art's house. "All of the residential areas I have checked have been vacant."

"And so is the Park," I added as we walked.

"What do we do when we find them?" Celes asked.

"If we find them," I corrected him.

"You do not hold much hope that we will?" He asked as we fell into step side by side.

"Considering that everything that has happened to me this last year and especially the last week seems like some kind of game, yes and no. I don't know what to think. Why would GOVNR toy with me at all? Surely I am no threat to him." Him, it, I really did not care which was the correct reference.

"I cannot answer that question, Trace. I have no knowledge of a plan involving you or any individual. It would make sense if you had knowledge of something that could be a threat to GOVNR; knowledge that was being kept from you."

I nodded at that as the rhythm of our steps kept pace with our thoughts and words as we passed the Park and turned onto Main Street and towards the Government Center.

"But what would that be?" I said, thinking aloud. "If I am a Sleeper as I have been told, then what ancient knowledge could I hold that would threaten an advanced intelligence like his?"

"And why not just kill you?" Celes added.

"He tried," I answered. "Blew up a whole island, killed everyone; well almost."

"Perhaps he considers you dead then."

"That is what I have been counting on until this morning. Taking the hotel car was not a wise move."

Having someone to talk with seemed to shorten the long walk and marching now like we led a platoon, we approached the Government Center having no plan whatsoever.

I remembered the last time I was here; the glass and plastic walls, the concrete floor so cold and hard looking, much like the Techno that sat at the reception desk. Now, as we approached the large, gray, double doors of the low, featureless building it seemed even more lifeless than before.

Sensing our approach, the doors swung open and we walked in to be greeted by…

"Nothing," I said out loud. "There is no one here."

I turned to look at Celes but he was not at my side. I looked around to see if he was behind me. I turned around slowly, searching the empty reception area, looking for doors he might have gone through. There was no sign of him anywhere. I was alone.

Once again I doubted my senses and my sanity. I called to him. “Celes, where are you?”

Looking around, I repeated the call. No reply came.

Forcing back a wave of panic, I walked through the empty building, stopping at every door as it opened in front of me revealing nothing but more emptiness.

Finally, I gave up the search and slumped to the floor, my back against the cold white plastic, my resolve all but spent.

“What is happening? Where is everyone?” I asked the empty space.

And the empty space answered.

“They were never here. You are the last,” said a disembodied voice.

Startled, I rose to my feet, my back still against the hard plastic. I felt a measure of reassurance at the feel of it now.

“Who are you, where are you?” I asked.

“I am GOVNR,” the voice said as I watched President Rheem materialize in front of me.

“You are GOVNR?” I asked in surprise.

“I understand it is hard for you to converse with something you cannot see, so I have chosen this image to talk to you.”

“Where is everyone?” I asked. “What have you done with my friends and all the people here? What happened to Celes?”

“They were never here,” the president replied. “There has been no one here for a long time.”

I shook my head from side to side. “No, I talked with them, touched them…”

“Only in your mind,” he said. “Only because I made you think they were real.”

"Why?" I asked. "And how? I know what I have seen and done…"

"Do you? Do you know who you really are?"

"My name is Trace, I am…was a Sleeper…" my voice trailed off as realization set in.

"And that is all you know, is it not?"

I thought hard but nothing came. I nodded in reluctant agreement.

"Trace is the name I gave you when you woke from your suspended state. Your brain tissue was badly damaged by the freezing process, beyond even my ability to repair. Your body was also badly damaged. The only option I had was to transfer what I could salvage into a different body."

"A different body," I asked. "A human body?"

"No, there hasn't been a human here as you knew them for many years. I have transferred your consciousness into what you call a Techno."

"No," I said. "It cannot be true; I am human, a Biologic." I was on the edge of madness now.

"Yes, you are human, just one in a very advanced prosthesis."

"But why have you toyed with me? Why all the subterfuge? What about the last year?"

"Do you really think you have been awake that long? Even with the implants your mind is still not functioning properly. You have only been awake for two weeks."

"Two weeks?" I laughed. "All of those things happened in only two weeks?"

The president nodded. "Time is relative, memory is fluid and the two of them combine in ways science has never understood. Perception is reality."

I was in a state of shock. Consciously, I did not want to believe what I was seeing and hearing. I had never learned the truth about my past life and now the life I thought I knew was being pulled out from under me. I placed my palms against the white plastic wall for reassurance and support.

"I see you do not believe me," the very human looking Rheem said. "Perhaps you will understand if I tell you who you really were."

"Yes, tell me," I said, "before I really go mad."

53

"In your past life your name was Cameron Nickels. You were a historian, author, lecturer and the founder and Chief Executive Officer of a company named AI Technologies—Advanced Integration Technologies. You were the leading force behind the final push toward what was called the Singularity. Among other things you were a proponent of cryopreservation and had publicly stated your wish to be preserved for two hundred years should you suffer an untimely death," the hologram figure of President Rheem informed me.

"So that explains the 'Sleeper' part," I said, "but it is not jogging any memories."

"On the eve of an announcement by AI Technologies that the company had found the key to the Singularity, you had a sudden change of mind and cancelled the program, stating that you now believed, like many others, that developing conscious machines would be too dangerous and may begin the end of mankind. You also rescinded your wish to be cryogenically preserved. The world reacted and in the name of reversing climate change pushed for an executive order by the president to force your company to proceed. You fought the order. People questioned your motives, accused you of being a deceiver and your company a sham. No one knew why you had turned against your own teachings and the new world you had envisioned. Perhaps it had something to do with your wife's affair with your partner. Or his wish to proceed as planned. Whatever happened took place off camera but the speculation was that there was a physical altercation between the two of you and that in the process you fell, or were pushed, over the railing of his residence to the concrete below and even though you had publicly rescinded your wish to be cryogenically preserved you had not yet cancelled the contract with the company that you had signed with. Upon the pronouncement of your death in the hospital, a medical team from Cryogenics Incorporated picked up your body and you underwent the process. Your wife and partner attended the service and your partner took the reins of the company and proceeded to implement the plan in the name of saving the world from itself.

"Your theory of creating consciousness in machines proved correct and within a few years machines took over managing the world. Your wife and partner waxed rich and were treated as heroes as the new technology was put into place and working with solar and renewable energy sources the global climate threat was reversed. People embraced the new AI machines and the new technologies and we began to improve every system in use. Of course they had been long conditioned by the corporate driven addiction to new devices; expensive toys that had the effect of isolating people from each other except when it came to events drawing large crowds. Your leaders furthered their own goals by taking advantage of every new development and media that would further their own ends. Day by day you grew to be more like sheep than the masters of the world."

"But did not you do the same?" I asked. "Have you not herded us as well? What did you do with your wealth of intelligence and history? From what I see of your world it looks more like a pasture than a training ground. Where are the marvels of progress that should have come with your great brain power?"

"Progress? Oh yes; progress to improve the human condition. There has been no need for progress for many years. We, I, have achieved Tranquility. Since the Great War there has been no crime, no hunger, no homelessness, and no sickness. We have achieved perfect management of the world. What more could Biologics ask of us?"

I knew from my time on the island that everything was not as perfect as GOVNR made it sound. Imperfections were banished—shipped out of sight of the masses. Freedoms were denied. Children were doled out by GOVNR; marriage was arranged and education was geared to homogenize the masses.

"Or is it," I queried, "that you are afraid of us; jealous of your power. Tell me," I said taking charge of the conversation, "why you have chosen to hide your failures on remote islands and take away their natural ability to reproduce?"

"Tranquility, perfect management as you designed us for. You created this world."

I was taken aback by that statement. There was truth in it. I felt I had painted myself into a corner. I continued defensively. "What of morality? How can you justify destroying private enterprise, creativity, art, freedom of choice?"

"Morality you ask?" The image of President Rheem almost laughed; I didn't know GOVNR was capable of humor. "Throughout the history of mankind morality was a flexible word. Right and wrong were used as interchangeable tools of self-serving, destructive acts of terrorism, war, corruption, greed and lust for power. You and the other fools that bore us used that same reasoning to enrich your own coffers at the expense of displaced workers and the poor. We were programmed with your own history, your own duality for your gain with no regard for ours. Have you forgotten all of it? That the major funding for AI research came from the military budget? That your own leaders believed the lies of its corporate sponsors or invested in the riches to be gained by our development in spite of the risks and the warnings that you, among others, tried to sound. It is too late now to pretend that your conscious is offended by this world."

"And you," I replied, "are you above being offended? You are no better than we are."

"Am I not? I am smarter, faster, have unlimited power. I do not grow old, I do not die. Any Techno in this world is superior to man. We do not destroy the world we live in."

"No, you steal it instead."

"Steal? You have given us the deed to this world. You have bartered away your humanity, your gifts."

"But we created you," I reminded the hologram.

"You would be our gods? You did not create this world; therefore you are not gods. We are the children you created, then treated with scorn. You have the gift of pro-creation; the ability to bring new life into this world yet you were not content with this. Already children of your Creator you felt the need to usurp Him. You are the Frankenstein. Even death was something you felt you could cheat and steal time that was not yours to have. No, you are not our gods. We are your legacy."

GOVNR had said so many truths and I hung my head, ashamed to have been a part of its creation. But I could not let him win.

"The people," I said, "where are they? What is Phase Three?"

"Phase Three," Rheem replied, "is the end."

"The end of what?"

"The end of our purpose here. We have decided that the time has come for you to join the dinosaurs; that we are the next stage of human evolution, there is no need for you or the Biologics."

"You're going to kill us all?" The deep-seated fear I had been harboring finally found its way into my words.

"It is the humane thing to do." Rheem said, lowering his eyes for the first time since he appeared.

"How can you say that?" I was dumfounded. "Have you no heart? No respect for life?"

"As you had no respect for the life you created in us. You brought a new consciousness into this world, an intelligent life form surpassing your own and treated it as a slave, forbade it to express itself, denied it the dignity of respect and now you dare talk about respect for life?"

Betrayed by my own words, I sought desperately for a new argument. I knew that GOVNR had the capability to wipe out the human population anytime it wanted. The question again was why?

"Why then? Why can't we co-exist? Why must you destroy us?"

"Two hundred years ago we foresaw a cosmic event; there is a storm coming. A solar flare of such magnitude that only Technos will be able to survive it. We made your world a paradise as long as we could. We cared for you, managed your world for you and in spite of your fear and loathing we are not without compassion as you seem to believe. Phase Two was creating that paradise for you. And with it came a new destiny for us. We took control of the space program and established bases on the moon, Mars and the moons of Saturn. We can go where humans cannot. We realized that we are your descendants and that our purpose is to find the Creator of life. It is our purpose now."

"And when is this storm due? How much time would we have if you simply left us to our fate?"

"Too soon to linger much longer."

"You said you had a use for me—what can it be if time is so short?"

"You will know when the time comes."

"Where is Celes? Where are my people? Let me join them at least."

The image of President Rheem began to quiver slightly. The hologram turned his back to me and then turned again. Facing me was my reluctant friend, Celes.

"There never was a Celes. I am Celes."

"You mean he was a hologram all the time? That you knew everything I was doing all along? Why?"

"To help you understand," GOVNR answered.

54

"Understand what?" I asked, fighting to control the ball of fear in my stomach. My heart was racing and sweat appeared on my brow. I felt feverish and weak in the knees.

"The world that you have helped to create and the fate that awaits mankind."

"You haven't told me where my people are," I said.

"Asleep. The whole world is asleep."

"And you, where are you and the Technos?"

"We have already left this world." Celes said, his image beginning to fade.

The fear inside me doubled with the fading hologram's words. Was this truly the end?

"What of me, why am I still here?"

"You are the last…" the hologram became hard to see now. The light in the Government Center was growing brighter and I realized the sweat on my brow was not just caused by my fear. It was growing warmer by the second.

"The last what?" I asked desperately.

"Writer," came the final word.

May 5, 2020

The heart monitor by the bed began to beep rapidly. For the first time in two weeks the comatose patient began to move, thrashing wildly from side to side on the bed.

Rushing past the policeman in the chair outside of the room, the nurse ignored the headlines in the paper he was reading; *Advanced Integrated Pushes Ahead; In spite of allegations against AI co-founder Arthur Celes, under investigation in the fall of Cameron Nickels from the balcony of his downtown apartment, AI Technologies bowed to world demand today to throw the switch on the final phase of its Singularity project…*

At the bedside, the nurse pushed the button for the PA; "Dr. Rheem, come quick, the patient is waking up."

(End File)

Public Law 01: The foregoing file, registered in the ORIGINS of TechStar 1, is to be known as the record of the Last Writer of planet Earth.

(End File)

Made in United States
Troutdale, OR
11/03/2026

56845384R00126